All-Russian Conference on Challenges for Development in Mining Science and Mining Industry Devoted to the 85th Anniversary of Academician Mikhail Kurlenya 2016

IOP Conference Series: Earth and Environmental Science
Volume 53

Novosibirsk, Russia
3 – 6 October 2016

Editor:

Mikhail Kurlenya

ISBN: 978-1-5108-3663-1
ISSN: 1755-1307

Printed from e-media with permission by:

Curran Associates, Inc.
57 Morehouse Lane
Red Hook, NY 12571

CURRAN ASSOCIATES INC. proceedings.com

Printed by Curran Associates, Inc. (2017)

For permission requests, please contact the Institute of Physics
at the address below.

Institute of Physics
Dirac House, Temple Back
Bristol BS1 6BE UK

Phone: 44 1 17 929 7481
Fax: 44 1 17 920 0979

techtracking@iop.org

Additional copies of this publication are available from:

Curran Associates, Inc.
57 Morehouse Lane
Red Hook, NY 12571 USA
Phone: 845-758-0400
Fax: 845-758-2633
Email: curran@proceedings.com
Web: www.proceedings.com

Table of contents

Volume 53

All-Russian Conference on Challenges for Development in Mining Science and Mining Industry devoted to the 85th anniversary of Academician Mikhail Kurlenya

3–6 October 2016, Novosibirsk, Russia

Accepted papers received: 13 January 2017
Published online: 9 February 2017

Preface

011001
OPEN ACCESS
All-Russian Conference on Challenges for Development in Mining Science and Mining Industry devoted to the 85th anniversary of Academician Mikhail Kurlenya

011002
OPEN ACCESS
Peer review statement

Papers

012001
OPEN ACCESS
Evolution of stress state of granular medium under multiple dynamic impacts

AP Bobryakov, VP Kosykh and AF Revuzhenko.....1

012002
OPEN ACCESS
Energy flows in rock mass under tidal deformation

SV Klishin and AF Revuzhenko.....5

012003
OPEN ACCESS
Non-Archimedean mathematical analysis methods in description of deformation of structurally inhomogeneous geomaterials

SV Lavrikov, OA Mikenina and AF Revuzhenko.....8

012004
OPEN ACCESS
Continuous monitoring of deformations in mines using laser interferometer–deformograph

MD Parushkin and DO Tereshkin.....15

012005
OPEN ACCESS
Features of stress state of support under varying displacement of free contour of underground excavation

VM Seryakov.....20

012006
OPEN ACCESS
Rock mass diagnostics based on microseismic monitoring data at Sheregesh deposit

TV Lobanova, GL Lindin, OL Trofimova, SK Shultaev and VV Prib.....24

012007
OPEN ACCESS
Estimate of rock mass stability in surface–borehole mining of high-grade iron ore

AS Sammal, SV Antsiferov, PV Deev and SV Sergeev.....30

012008
OPEN ACCESS
Evolution of microseismicity parameters depending on geomaterial deformation stages

VI Vostrikov, OM Usoltseva, PA Tsoi and VN Semenov.....34

012009
OPEN ACCESS
Effects of structure on deformation and strength characteristics of transversely isotropic man-made geomaterials

OM Usoltseva, PA Tsoi and VN Semenov.....38

012010
OPEN ACCESS
The surface waves-based seismic exploration of soil and ground water

AS Serdyukov, AV Yablokov, GS Chernyshov and AV Azarov.....43

012011
OPEN ACCESS
Laboratory research of hydraulic fracturing with tangential loading of borehole wall

MV Kurlenya, AV Patutin, LA Rybalkin, SV Serdyukov and TV Shilova.....48

012012
OPEN ACCESS
Spatial discrete element analysis of problem on floor buckling in underground openings

SV Klishin and AF Revuzhenko.....52

012013
OPEN ACCESS
Evaluation of a new trend in the geotechnology for steeply dipping vein deposits in Kyrgyzstan

KCh Kozhogulov and ZhM Ganiev.....58

012014
OPEN ACCESS
Advancement in modern approaches to mineral production quality control

EV Freidina, AA Botvinnik and AN Dvornikova.....61

012015
OPEN ACCESS
Geotechnical approaches to coal ash content control in mining of complex structure deposits

SA Batugin, VL Gavrilov and EA Khoyutanov.....66

012016
OPEN ACCESS
One of the current trends of geotechnology modernization

SM Tkach and SA Batugin.....71

012017
OPEN ACCESS
Effect of georesource–consumer process flows on coal loss in energy supply of the Polar regions in Yakutia

SM Tkach and VL Gavrilov.....77

012018
OPEN ACCESS
The problems and prospects of the public–private partnership in the Russian fuel and energy sector

SM Nikitenko and EV Goosen.....82

012019
OPEN ACCESS
Technology and means of a coal seam interval hydraulic fracturing for the seam degassing intensification

VI Klishin, GY Opruk and AL Tatsienko.....91

012020
OPEN ACCESS
Structural change and economic growth in modern Russia: The role of "resource-type" regions

SN Levin, DV Kislitsyn and KS Sablin.....98

012021
OPEN ACCESS
Development of chemical compositions for impervious screens in rocks

MV Kurlenya, SV Serdyukov, TV Shilova and AV Patutin.....109

012022
OPEN ACCESS
Application prospects for impact machines in the area of strengthening of joint welds

AD Abramov and MS Galay.....114

012023
OPEN ACCESS
Layout and variables of air-powered drive of small-size pneumatic winch

VP Gileta, AV Baris and YuV Vanag.....117

012024
OPEN ACCESS
Design of abrasive tool for high-rate grinding

AS Ilinykh.....121

012025
OPEN ACCESS
Determination of variables for air distribution system with elastic valve for down-the-hole pneumatic hammer

AYu Primychkin, AS Kondratenko and VV Timonin.....125

012026
OPEN ACCESS
Straight hole drilling machines for coal mines

VV Timonin, DI Kokoulin, SE Alekseev and B Kubanychbek.....129

012027
OPEN ACCESS
Causes of wear of PDC bits and ways of improving their wear resistance

VV Timonin, AS Smolentsev, I O Shakhtorin, NI Polushin, AI Laptev and AS Kushkhabiev.....133

012028
OPEN ACCESS
Simulation modeling and tracing optimal trajectory of robotic mining machine effector

VN Fryanov and LD Pavlova.....139

012029
OPEN ACCESS
Borehole tool for studies in coalbed degasification wells

SV Serdyukov, AV Patutin and TV Shilova.....146

012030
OPEN ACCESS
Unbalanced rock mass vibration generator and its bench testing

LA Rybalkin, AV Azarov and AS Serdyukov.....150

012031
OPEN ACCESS
Main development trends and some technical decisions on mining tools equipped with super-hard composite materials inserts

LT Dvornikov, PD Krestovozdvizhensky, SM Nikitenko, VA Korneyev and PA Korneyev.....156

012032
OPEN ACCESS
Geological and technological evaluation of gold-bearing mineral material after photo-electrochemical activation leaching

DV Manzyrev.....166

012033
OPEN ACCESS
Role of hydrogen ions in standard and activation heap leaching of gold

YuI Rubtsov.....171

012034
OPEN ACCESS
Cluster chemical reactions at mineral–liquid interface in metal leaching by photo-electroactive water-and-gas emulsions

AG Sekisov.....176

012035
OPEN ACCESS
Small-scale and large-scale testing of photo-electrochemically activated leaching technology in Aprelkovo and Delmachik Mines

AG Sekisov, AYu Lavrov and YuI Rubtsov.....180

012036
OPEN ACCESS
Explosion-assisted preparation of dispersed gold-bearing different-grade ore for selective mining

AI Trubachev and NV Zykov.....183

012037
OPEN ACCESS
Preparation of ore blocks for mine leaching by reagent explosion injection

YuS Shevchenko.....185

012038
OPEN ACCESS
Relaxation phenomena and jointing in ore bodies under explosive injection treatment

YuS Shevchenko.....188

012039
OPEN ACCESS
Evaluation and development of integrated technology of rare metal concentrate production in high-level ore processing at Zashikhinsk deposit

MS Khokhulya, TN Mukhina, V A Ivanova, G V Mitrofanova, A V Fomin and VD Sokolov.....191

012040
OPEN ACCESS
Environmental pressure reduction with a new method of noble metal recovery

EV Filippova.....197

IOP Conf. Series: Earth and Environmental Science **53** (2017) 011001 doi:10.1088/1755-1315/53/1/011001

2016 Conference Proceedings

Mining Science: Basics and Applications

October 3–6 2016

Novosibirsk, Russia

It is a pleasure for us to have this opportunity and offer our foreign readers the selected papers presented at the annual 2016 Conference on Mining Science: Basics and Applications. This year the Conference has been devoted to the 85th Anniversary of Mikhail Kurlenya, Academician of the Russian Academy of Sciences.

The Conference has been hosted by the Institute of Mining of the Siberian Branch of the Russian Academy of Sciences under auspices of the Presidium of the Siberian Branch RAS and the Scientific Council on Mining of the Russian Academy of Sciences.

The Conference has become a forum of academic scientists and practitioners from Russia's top mining companies such as Norilsk Nickel, Evrazruda, ALROSA, ELSIB and Siberian Association for Industrial Hydraulics and Pneumatics.

The Conference provided a platform for the discussion of the latest theoretical and experimental research findings in the field of new mining technologies and in the area of mining machines and automated equipment engineering for enhancement of efficiency of mineral extraction and processing and special construction work. The scope of the Conference encompassed such topics as:

- Geomechanics and induced geodynamics
- Advances in geotechnology and renovations in hybrid open pit/underground mining
- Principles of machine engineering for mining and underground construction
- Fundamentals and modern development in integrated utilization of rebellious minerals and wastewater treatment

Mikhail Kurlenya

Academician RAS

Editor-in-Chief

Peer review statement

All papers published in this volume of *IOP Conference Series: Earth and Environmental Science* have been peer reviewed through processes administered by the proceedings Editors. Reviews were conducted by expert referees to the professional and scientific standards expected of a proceedings journal published by IOP Publishing.

Evolution of stress state of granular medium under multiple dynamic impacts

AP Bobryakov, VP Kosykh* and AF Revuzhenko**

Chinakal Institute of Mining, Siberian Branch, Russian Academy of Sciences, Novosibirsk, Russia

E-mail: *v-kosykh@yandex.ru, **revuzhenko@yandex.ru

Abstract. The paper describes experimental measurement of pressure in marble chips placed in cylindrical reservoir subjected to weak impacting within a long time. It is found that pressure in the granular material changes non-monotonically but fluctuates relative to hydrostatic pressure with the amplitude equaling order of magnitude of the latter. The fractal analysis of the experimental curves of pressure and number of impacts is performed. The Hurst exponent ranges as 0.67–0.73, and the process of pressure variation is fractal and possesses long-term "memory."

It is common knowledge that geomaterials—rocks, soils, granular media—have various kinds of equilibria. This means that having barely the information on body forces and boundary conditions, it is impossible to assess stresses inside a domain. On the other hand, stress assessment is the basic objective of geomechanics. The question is that there can be self-balancing stresses in a medium [1–3]. When solving a linear problem on secondary stresses, it is allowed to neglect initial stresses. However in case that local damage or local transition to limiting state is probable, the role of the initial stress field may be critical, for instance, in a problem on weak impacts exerted on a medium. Experiments show that such impacts induce irreversible deformations in the medium. So, a medium evolves from one "intrinsic" state to another.

This study focuses on the effect of weak but long-term impacts on a medium.

Which impacts can be assumed weak? Initially, a granular medium has only one characteristic of the dimension of stress. This is the cohesion k. And, when a medium is in the limiting state, it has one more characteristic—internal friction angle φ. The latter is dimensionless and is included in the limiting state conditions together with the uniform compression:

$$\tau = -\sin\varphi \cdot \sigma + k\cos\varphi .$$

Impacts may be characterized by a pressure p^*. This allows a natural dimensionless combination of stresses

$$\lambda = \frac{p^*}{-\sin\varphi \cdot \sigma + k\cos\varphi} .$$

Accordingly, "weakness" of an impact is conditioned by a specific element of a medium, and by strength characteristic and stress state of this element. Consequently, an explosion load is high in the near and medium-range zones and is weak in the rest (major part) of rock mass.

A granular medium is composed of many discrete solid particles that can carry out translational or rotational movements under certain energy input. This motion can include both differential movements of individual particles and coordinated movement of particle clusters temporarily composed and decomposed. The detailed research has shown that particles in a granular medium under deformation make clusters (actually, blocks) [4]. Deformations mainly occur at the boundaries of the clusters and are comparatively low inside the clusters. Clusters of particles with flat faces are better observed. By way of illustration, the squashing test of a bunch of six-sided pencils tied up by an elastic rubber from outside can be presented [5].

The evolution of stress state of a specimen composed of marble chips was studied in the experimental treatment of the specimen by weak impacts within a long time (see Figure 1). A steel cylindrical glass 1 with a diameter of 100 mm and 220 mm high, with the wall thickness of 0.3 mm was filled with marble chips up to the top using a funnel. Two sensors 3 were embedded in marble chips specimen to measure vertical pressure. The sensors were $12\times25\times3$ mm in dimension. The sensors were buried at a depth of 150 and 110 mm from the specimen top surface. The glass was fixed on a stiff basis. An impacting tool was arranged nearby. The nose-piece of the impacting tool was made of rubber. The distance between the glass bottom and the impact point was 60 mm.

To avoid temperature effect on the sensors, the test system was placed in a thermostat. As soon as the system was heated and the sensors gave steady-state reaching, the impacting tool was actuated. The impact energy E was constantly $3.85\cdot10^{-3}$ J during the test and the blow frequency was 80 impacts per minute. Sampling period was 10 s. Particles of marble chips were 1 mm in size on average.

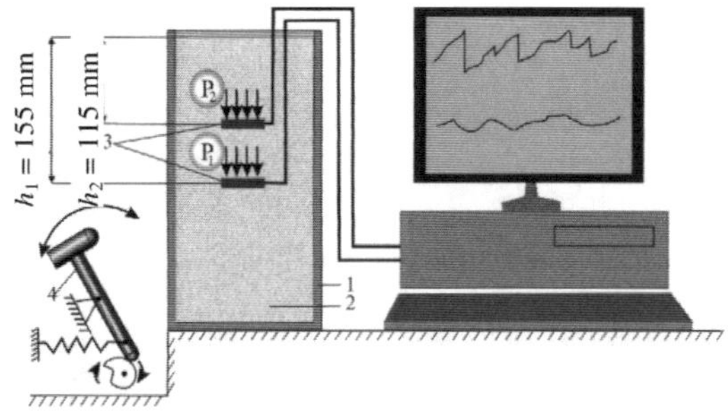

Figure 1. Schematic of the test: 1—cylindrical glass; 2—marble chips; 3—normal stress sensors; 4—impacting device.

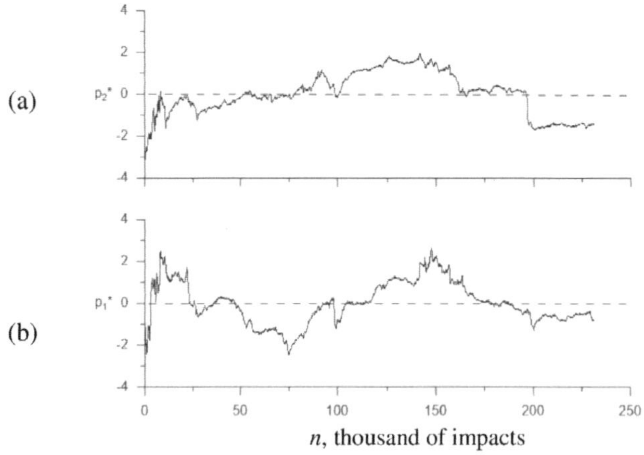

Figure 2. Normalized plots of vertical pressure in marble chips package versus multiple weak impacts: (a) upper sensor; (b) lower sensor.

The pressure–impact curves plotted based on the test data were processed. At the first stage, linear trends adjusted to the experimental curves using the least square method were withdrawn from the plots. The second stage of the processing was normalization of the plots through dividing of pressure readings by standard deviation. The resultant normalized curves are presented in Figure 2.

Evidently, the pressure curves in Figure 2 are non-monotonic. There are peaks and troughs. The same pressure pattern was experimentally obtained in quartz sand tests [6], and the tests showed that stress deviation from uniform compression could reach 20–150% in the course of the experiment.

By this time, researchers have found that many time series, although looking random, are not random actually. Using many model and natural series, it has been shown that they often contain information on begetting mechanisms. For example, the R/S analysis reveals the fractal nature of a process and the process "memory" [7–9]. The R/S analysis was introduced in the practice of the empirical analysis of time series by hydrologist Harold Edwin Hurst in 1960s. He discovered that for many experimental time series, the ratio R/S grows exponentially with an increase in the series length L

$$(R_L / S_L)^H / L \sim \text{const}(L),$$

where R_L and S_L are, respectively, the series amplitude and the standard deviation dependent on the series length L; H is the parameter of the self-similarity of a time series (Hurst exponent).

For a time series X_t at $t = 1, 2, 3, \ldots L$, the amplitude R_L is determined as the difference between the maximum and minimum deviations of X_t from the average X_0

$$R_L = \max[Y_t(k)\,|_{k=1..L}] - \min[Y_t(k)\,|_{k=1..L}],$$

where Y_t is the sum of the deviations from the current average X_0 over the interval from 1 to L

$$Y_t = (X_1 - X_0) + (X_2 - X_0) + (X_3 - X_0) + \ldots (X_k - X_0).$$

Hurst analysis is intended to find slope of the plot $R/S \sim \tau^H$ in a log–log frame, where R is the amplitude of the time series in the given time interval τ; S is the standard deviation in the same interval.

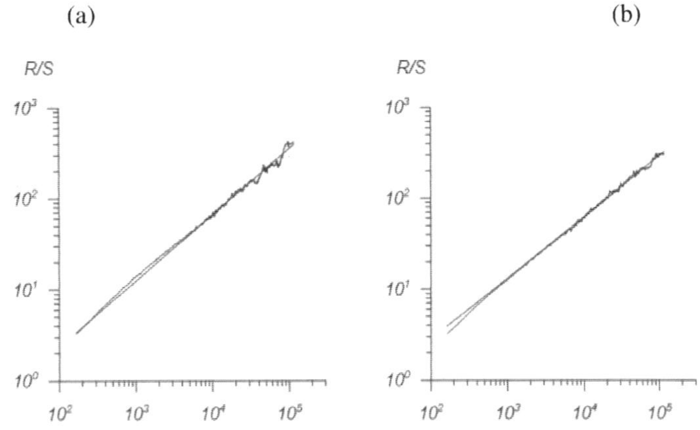

Figure 3. Normalized pressure amplitude versus the width of interval τ (number of impacts): (a) lower sensor; (b) upper sensor.

For a random (stochastic) process, with independent increments and finite dispersion, $H = 0.5$. Deviation of H from 0.5 takes place when the analyzed processes possess "memory" and is the evidence of fractal properties of these processes. When $H > 0.5$ time series have a sustainable tendency (persistency). When $H < 0.5$ the increase in the observed value changes to a decrease and vice versa (anti-persistency). The Hurst method is very stable and needs no a priory assumption on the character of distribution of values. For many natural processes, e.g. river flows, river and lake stages, amount of precipitation, thickness of bedded precipitation, thickness of growth rings, number sunspots etc., H range is 0.6–0.8 [7, 8].

The outcome of the R/S processing of the experimental pressure curves from Figure 2 is depicted in Figure 3.

It is seen that the curves of R/S and the interval width (number of impacts) τ are well approximated by exponential functions. This implies the fractal nature of the studied process, the Hurst exponents for the lower and upper sensors are 0.73 and 0.67, respectively.

Thus, the time series under analysis have a stable tendency (persistency) while the process that induces them is fractal and possesses long-term "memory."

Apparently, this is connected with that the impulse given to particles under percussion depends on arrangement of the particles at any specific time. Mutual arrangement of particles is changed from blow to blow and, consequently, conditions of contact interaction between particles (orientation of interfaces, contact stresses) are altered, too. For this reason, it is possible to state that displacement of a particle under each impact is more or less random. The result of this process is continuous re-packing of material, with formation of stable clusters of hierarchical (self-similar) structure here and there. Under long-term impact, clusters are decomposed and originated again.

These research findings conform with the results obtained in [10–12], where t is shown that movement and deformation in granular media are characterized by fluctuations and jumps of pressure. Deformations localize and block structures similar to fault-and-block systems are formed. The fractal distribution of sizes of blocks formed in a granular medium under external impact was observed in [13].

References

[1] Moroz AI 2004 *Self-Stress State of Rocks* Moscow: MGGU (in Russian)
[2] Birger IA 1963 *Residual Stresses* Moscow: Mashgiz (in Russian)
[3] *Organizational Standard STO 36554501-019-2009 Identification of Self-Stress State in Rocks. NITs Stroitelstvo* Moscow 2010 (in Russian)
[4] Kosykh VP 2012 Experimental investigation in variable density of granular material under deformation cycling *Journal of Mining Science* Vol 48 No 6 pp 990–995
[5] Lavrikov SV and Revuzhenko AF 1991 One experimental rock model *Journal of Mining Science* Vol 27 No 4 pp 288–293
[6] Kazantsev AA Kosykh VP and Revuzhenko AF 2015 Spectral analysis of granular material reaction to long-term weak dynamic effect, *IOP Conf. Series: Materials Science and Engineering* 91 012089 doi:10.1088/1757-899X/91/1/012089.
[7] Feder J 1988 *Fractals* Springer US
[8] Schroeder M 2009 *Fractals, Chaos, Power Laws* Dover Publications
[9] Lukk AA, Deshcherevsky AV, Sidorin AYa and Sidorin IA 1996 *Variation of Geophysical Fields as the Deterministic Chaos in Fractal Media* Moscow OIFZ RAN (in Russian)
[10] Leonov MG 2002 Tectonics of complex-dislocated areas in the Earth's crust (Problems and solutions) *Litosfera* No 1
[11] Behringer RP, Howell D, Kondic L, Tennakoon S and Veje C 1999 Predictability and granular materials *Physica D* Vol 133 Nos 1–4
[12] Revuzhenko AF 2006 *Mechanics of Granular Media* Springer-Verlag Berlin Heidelberg
[13] Mcdowell GR, Bolton MD and Robertson D 1996 The fractal crushing of granular materials *J. Mech. Phys. Solids* Vol 44 No 2

Energy flows in rock mass under tidal deformation

SV Klishin[*] and AF Revuzhenko

Chinakal Institute of Mining, Siberian Branch, Russian Academy of Sciences, Novosibirsk, Russia

E-mail: [*]sv.klishin@gmail.com

Abstract. Under analysis is the stress state of an elliptical domain under varying loading conditions. The energy flow lines are plotted. The paper demonstrates the effect of the boundary conditions on the shape of the flow lines.

Tidal deformation of the Earth is a global process. It exerts influence, in a varying degree, on all geological processes on the planet: evolution of stresses in the interior of the Earth, including mineral mining depths, mass transfer, deformation of blocks on different scales, behavior of interface areas between the blocks, etc. [1]. Investigation and understanding of the tidal deformation of the Earth is an interesting challenge at this time.

Tidal deformation is connected with the energy physics. On the whole, tidal deformation is induced by kinetic energy of rotation of the Earth. Transformation of the energy of rotation into the energy of deformation of the solid and viscous Earth results in deceleration of Earth's rotation and in elongation of the day length. In kinematic models of tides, energy flows through the boundary of a body. This energy, in a specified approximation, ensures the same kinematics of mass movement as in deformation under body forces.

It seems interesting to address energy flows in these situations. Let the Cartesian coordinate system Oxy contain an elliptical domain with a center (0, 0) and axes oriented along the axes Ox and Oy

$$\frac{x^2}{(1+m)^2}+\frac{y^2}{(1-m)^2}\le 1\ , \qquad (1)$$

where $0 \le m << 1$.

Let us compare deformation of a perfect granular medium and a uniform viscous fluid. According to [2], when Keplerian velocities are assigned at the boundary of an elliptical domain, velocity distribution inside the domains is always linear. And this fact is independent of rheology of a medium. Deviation of boundary conditions from a Keplerian kind results in nonlinearity of velocity field and, consequently, to the velocity dependence on rheology of the medium. The basic characteristics of a nonlinear field are observable in the models of a perfect granular medium and Newtonian viscous fluid. The both models in a first approximation with respect to m << 1 yield the same kinematics. On this basis, we analyze the stationary solutions of the Navier–Stokes equations

$$\nu\Delta u-\frac{1}{\rho}\frac{\partial p}{\partial x}=u\frac{\partial u}{\partial x}+v\frac{\partial u}{\partial y},\quad \nu\Delta v-\frac{1}{\rho}\frac{\partial p}{\partial y}=u\frac{\partial v}{\partial x}+v\frac{\partial v}{\partial y},$$
$$\frac{\partial u}{\partial x}+\frac{\partial v}{\partial y}=0, \qquad (2)$$

where Δ is the Laplace operator; $u(x, y)$, $v(x, y)$ are the components of the velocity vector $\boldsymbol{v}$; $p(x, y)$ is the pressure; ρ is the density; ν is the viscosity coefficient. The boundary conditions for the elliptical domain (1) are set as

$$|\boldsymbol{v}| = v^0 = \text{const}; \ \boldsymbol{v} \cdot \boldsymbol{n} = 0. \tag{3}$$

Here, $\boldsymbol{n}$ is the vector of the outer normal drawn to the boundary. With the scheme from [3], the solution to the problem (2) with the boundary conditions (3) is given by:

$$u(x, y) = -y + m\left(-3y + 2y^3\right) + m^2\left(\frac{7}{4}y + \frac{7}{2}y^3 - \frac{9}{4}y^5 - \frac{21}{2}x^2 y + \frac{15}{2}x^2 y^3 + \frac{15}{4}x^4 y\right),$$
$$v(x, y) = x + m\left(-3x + 2x^3\right) + m^2\left(-\frac{7}{4}x - \frac{7}{2}x^3 + \frac{9}{4}x^5 + \frac{21}{2}xy^2 - \frac{15}{2}x^3 y^2 - \frac{15}{4}xy^4\right). \tag{4}$$

Particle transport in a medium and the mechanism of deformation are determined from the numerical integrating of the system of ordinary differential equations

$$\frac{dx(t)}{dt} = u(x, y), \ \frac{dy(t)}{dt} = v(x, y)$$

with the initial conditions

$$x(0) = x^0, \ y(0) = y^0.$$

Figure 1 depicts the trajectories of particles that are in the major semiaxis of the ellipse at a reference time for $m = 0.1$, 0.2 and 0.3. It is seen that all particles move along closed trajectories, and the periods of revolution around the center are different along different trajectories. Because of this difference, internal deformation of the body grows infinitely with time. According to [2], this effect can be interpreted as the directional mass transfer. In this way, the proposed method of loading can be considered as specification of a definite wave flow on a body surface.

(a) (b) (c)

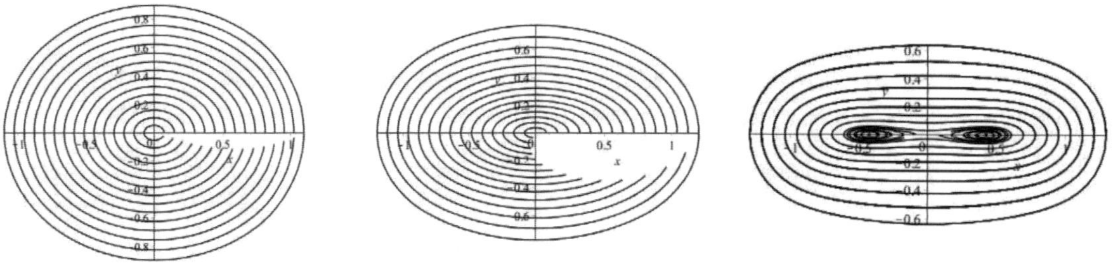

Figure 1. Trajectories of particles at (a) $m = 0.1$; (b) $m = 0.2$; (c) $m = 0.3$.

Plotting of lines of energy flow reduces to finding a vector field [4–6]

$$\boldsymbol{E} = -\{\sigma_{xx}u + \sigma_{xy}v, \sigma_{xy}u + \sigma_{yy}v\} \tag{5}$$

conditioned by stresses σ_{xx}, σ_{yy}, σ_{xy} and velocities u and v.

It follows from the analysis of deformation processes in different media that stresses in an arbitrary element of a body depend on forces and displacements pre-set at the entire boundary of the body. On the other hand, energy flows into the body not from the entire boundary but from its specific sectors. For the problem under consideration, energy glow lines at different m are shown in Figure 2. It is seen that at small m, the energy flow lines envelope the origin of the coordinate system and shape a narrow region of confluence at the intersections of the boundary and major semiaxis of the ellipse. As m is increased, the flow gets closer to the Ox axis (Figure 2b), and the confluence boundary drifts toward the Oy axis. The further increase in m results in generation of an interface at which some lines of energy flow go back to the domain boundary while the other energy flow lines tend toward the coordinate origin.

(a) (b) (c)

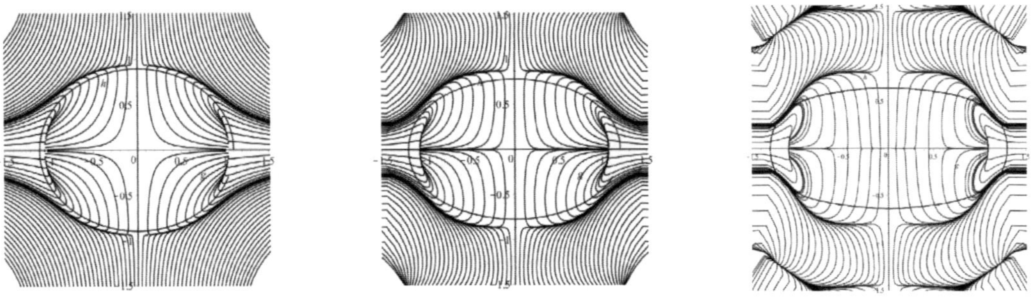

Figure 2. Energy flow lines for (a) $m = 0.1$; (b) $m = 0.2$; (c) $m = 0.3$.

Figure 3 shows the absolute vectors $\boldsymbol{E}$ for different m. In all the cases, the absolute vector (5) is close to zero in the vicinity of the coordinate origin.

(a) (b) (c)

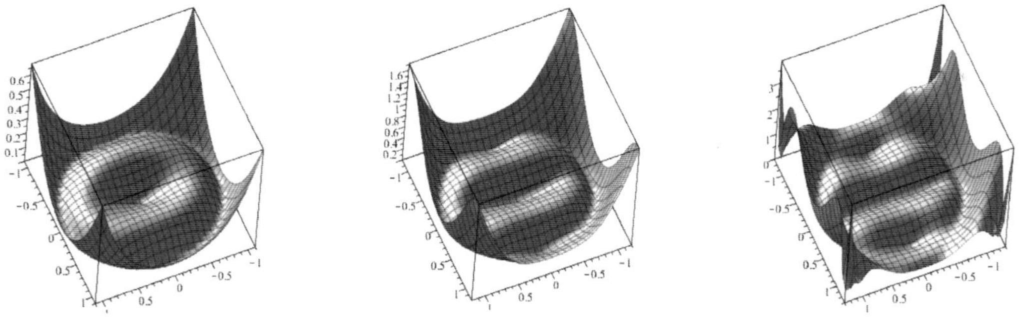

Figure 3. Absolute vectors of energy flow for (a) $m = 0.1$; (b) $m = 0.2$; (c) $m = 0.3$.

Conclusion

During deformation of a solid, energy flow in an elastic material is similar to the compressible fluid flow, and it is always possible to identify segment of the body boundary where energy flows in the body. The kinematic model of tidal deformation discussed in this paper illustrates that energy inflow in the domain induces transport of internal mass.

References

[1] Adushkin VV and Spivak AA 2010 Tidal force as a trigger of geophysical processes in the ambient medium *Trigger Effects in Geosystems* Moscow: GEOS (in Russian)

[2] Revuzhenko AF 2013 *Tidal Waves and Directional Mass Transport in the Earth System* Novosibirsk: Nauka (in Russian)

[3] Muskhelishvili NI 1966 *Some Basic Problems of Mathematical Theory of Elasticity* Moscow: Nauka (in Russian)

[4] Umov NA 1950 *Selectals* Moscow-Leningrad: Gostekhizdat (in Russian)

[5] Love AEH 1944 *A Treatise on the Mathematical Theory of Elasticity* Dover Publications

[6] Revuzhenko AF and Klishin SV 2009 Energy flux lines in a deformable rock mass with elliptical openings *Journal of Mining Science* Vol 45 No 3 pp 201–206

Non-Archimedean mathematical analysis methods in description of deformation of structurally inhomogeneous geomaterials

SV Lavrikov*, **OA Mikenina**** **and AF Revuzhenko*****

Chinakal Institute of Mining, Siberian Branch, Russian Academy of Sciences, Novosibirsk, Russia

E-mail: *lvk64@mail.ru, **olgarev@yandex.ru, ***revuzhenko@yandex.ru

Abstract. Under analysis is an approach to mathematical modeling of structurally inhomogeneous rocks considering structural hierarchy and internal self-balanced stresses. The fields of stresses and strains at various scale levels of rock mass medium are characterized using the non-Archimedean analysis methods. It is shown that such model describes accumulationtion of elastic energy in the form of internal self-balanced stresses on a micro-scale. The finite element algorithm and a computer program are developed to solve plane boundary-value problems. The calculated data on compression of a rock specimen are reported. The paper shows that the behavior of plastic strain zones largley depends on the pre-set initital micro-stresses.

1. Introduction

One of the key problems of geomechanics is calculation of stresses and strains in a unit element of rock mass. The body forces and the boundary conditions are pre-set. In the meanwhile some part of internal stresses in a self-stressing rock mass can be mutually balanced. And boundary conditions offer no information on them in this case. So, it is required to know the body forces, the boundary conditions and the conditions of how the rock mass was formed. Some information on self-balancing stresses is obtained experimentally, using special testing procedures [2, 3]. These data are useful in selecting parameters for a mathematical model of self-balancing stresses. One of the modeling methods introduces hypotheses on microdeformation of a medium, then uses averaging, and for the averaged values the continuum equations are derived. This approach assumes that the unit volume of the medium is subject to uniform stresses and strains [4].

An alternative approach is based on using non-Archimedean variables. The model avoids the assumption of the uniform state of the unit volume. Equations of equilibrium and motion are formulated specifically for structural elements. The size of the elements is set in proportion to the value of $1/n$, where n is a sufficiently high natural number. Then, a discrete model is constructed. In the classical analysis, the next step is the analysis of the limit when $n \to \infty$ and, accordingly, $1/n \to 0$. As a result, the finite difference equations transform into the differential equations. In the non-Archimedean analysis, the number n tends not to infinity but to an actual infinitely high number ω. The size of a unit element tends to E. The value of E is smaller than any positive rational number but $E > 0$. For this reason, the finite difference equations transform not in the differential equation but in the actual infinitesimal difference equations. They are analyzable at a varying degree of accuracy, e.g. assuming that $E = 0$, $E^2 = 0$ or $E > 0$, $E^2 = 0$, etc. Sometimes, the replacement of

$1/n$ by E is of consequence. In the other situation, this replacement may of critical value. For instance, when the problem source parameters already include the values E, ω, or when new measurements of space and time on a macrolevel are included. Here we mean measurements included in the structure of a real number as a real number is not only a multiscale straight line segment of the length E but an infinite entity attributed with a diameter of the order of E (conclusion of Cantor's concept of number [5]).

2. Mathematical model

This paper describes the analysis of the model of self-stressing rock mass (Figure 1, plane strain) [6, 7]. Initially, elastic particles are at the points of a square lattice. Sliding between the particles is possible in accordance with the laws of plasticity, dry or viscous friction, or their combinations. Along the diagonal, the particles are connected by elastic elements (a porous medium is ether elastic or of Vickers type). Forces applied to the elastic elements can be balanced by the forces applied to contacts, consequently, internal self-balancing stresses can be considerable in case that external stresses are absent.

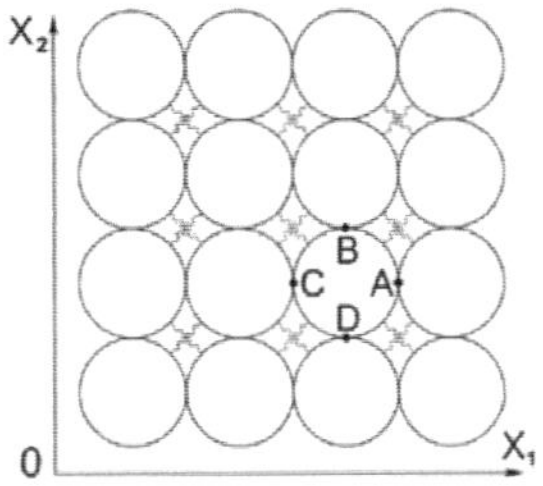

Figure 1. Model of a structurally nonuniform medium capable to accumulate elastic energy.

Any mathematical model should satisfy certain consistency constraints, e.g. to convert into model of the linear elasticity theory in a special case. This offers a way for further generalizations. Let $2L$ be the size of a domain subject to deformation and $2L/n$ –be the size of a structural element of the domain (particle). It is assumed that particles are linearly elastic, and there are no pores and sliding at contacts of the particles is absent. Furthermore, it is assumed that point moments cannot propagate through the contacts. In this case, there are four force vectors F and four displacement vectors u for each particle. One vector conforms with two scalar components (plane strain). Temperature and other parameters can also be added. The total amount of the scalar unknowns is: $4n(n-1)$—displacements at contacts, $4n(n-1)$—forces at contacts, $8n$—displacement at boundaries, $8n$—forces at boundaries. Thus, we have $8n^2+8n$ variables.

Regarding the equations, two equations of equilibrium and one equation of moments are fulfilled for each particle in the set n^2, i.e. $3n^2$ equations all in all. Then, at each of $4n$ boundary contacts, two conditions should be set (either forces or displacements, or combination), which gives $8n$ conditions altogether. A key point is the constitutive relations. The four points A, B, C, D agree with the four vectors of displacement—eight degrees of freedom. Constitutive relations may only include such combinations of degrees of freedom, which are independent of translation and rotation of a particle, which means that three degrees of freedom are to be withdrawn. Consequently, there are five invariant combinations of displacements. The forces also have eight degrees of freedom. The vector of sum of forces and the moment are to be zero. Thus, constitutive relations connect five invariant combinations of displacements and five force characteristics. There should be five constitutive relations for one particle and $5n^2$ constitutive relations for n^2 particles. This totals $8n^2+8n$ equations. Balance has converged: the system is closed and reduces to $8n^2+8n$ algebraic equations in terms of $8n^2+8n$ unknowns.

Now, a unit volume needs five constitutive relations. On the other hand, there are merely three constitutive relations for an elastic body. This means that the classical theory contains assumptions that match two equations, moreover, these equations are of the same significance as the equations involved in the Hooke law. Let us formulate them explicitly.

First, we select five invariant combinations of displacements for the constitutive relations. The number of decision is unlimited. We chose the closest variant to the linear elasticity

$$\begin{aligned} e_{11} &= \frac{\delta u_1}{\delta x_1} = \frac{u_1(A) - u_1(C)}{2l}, \\ e_{22} &= \frac{\delta u_2}{\delta x_2} = \frac{u_2(D) - u_2(B)}{2l}, \\ e_{12} &= \frac{1}{2}\left(\frac{\delta u_1}{\delta x_2} + \frac{\delta u_2}{\delta x_1}\right) = \frac{u_1(D) - u_1(B)}{4l} + \frac{u_2(A) - u_2(C)}{4l}, \end{aligned} \tag{1}$$

$$\begin{aligned} \kappa_1 &= \frac{u_1(C) + u_1(A)}{4l} - \frac{u_1(D) + u_1(B)}{4l}, \\ \kappa_2 &= \frac{u_2(D) + u_2(B)}{4l} - \frac{u_2(C) + u_2(A)}{4l}. \end{aligned} \tag{2}$$

Here, δ denotes increment within a particle. Now, it is possible to re-write the equations of the classical elasticity theory as

$$\begin{aligned} e_{11} &= \frac{1}{E}(\sigma_{11} - \nu\sigma_{22}), \quad e_{22} = \frac{1}{E}(\sigma_{22} - \nu\sigma_{11}), \\ e_{12} &= \frac{1+\nu}{E}\sigma_{12}, \quad \kappa_1 = 0, \quad \kappa_2 = 0, \end{aligned} \tag{3}$$

where E, ν–const; $\sigma_{ij} = F_{ij}/2l$, $\sigma_{12} = \sigma_{21}$, $i, j = 1, 2$. The first three equations are the Hooke law. The last two equations are not formulated explicitly in the classical theory, although it contains some information about them. Where? The classical elasticity theory postulates diffeomorphism [8]. In other words, it is assumed that all functions are sufficiently smooth. This means that any local functions can be represented by a linear function, for instance

$$\begin{aligned} u_1 &= a_{11}x_1 + a_{12}x_2, \\ u_2 &= a_{12}x_1 + a_{22}x_2, \end{aligned} \tag{4}$$

where $a_{11}, ..., a_{22}$ are constants. Placing (4) in (2) yields $\kappa_1 \equiv 0$, $\kappa_2 \equiv 0$. The converse is valid, too. Consequently, the wanted two equations from (3) are "hidden" in the postulate of diffeomorphism. Technically, this allows transition from "derivatives" within the limits of a particle, $\delta/\delta x_i$, to the ordinary derivatives $\partial/\partial x_i$, i.e. to classical theory. Accordingly, the consistency condition for the models under analysis is satisfied.

In case of no diffeomorphism, it seems relevant to use the non-Archimedean analysis representation

$$\begin{aligned} u_1 &= u_1(x_1, x_2, \xi_1, \xi_2), \\ u_2 &= u_2(x_1, x_2, \xi_1, \xi_2). \end{aligned} \tag{5}$$

Here, the fixed values x_1, x_2 are the centers of particles, and ξ_1, ξ_2 are the coordinates within a fixed particle. The nonlinearity of the functions with respect to the coordinates ξ_1, ξ_2 agrees with the case when $\kappa_1 \neq 0$, $\kappa_2 \neq 0$. In a general case of allowable sliding at contacts and eigen stresses, the situation will be similar. From the viewpoint of mechanics, the model in Figure 1 conforms with the parallel connection of elastic and plastic elements in Figure 2.

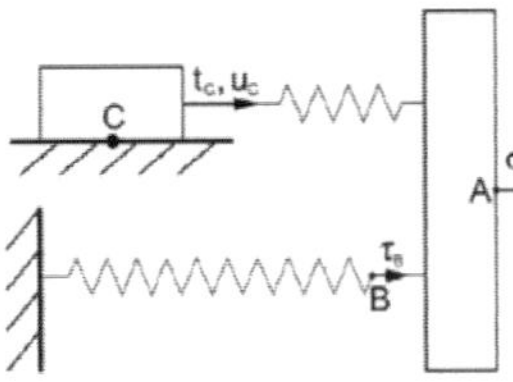

Figure 2. Mechanical model of a medium having structure.

The macrostress σ_A matches the stress τ_B in a porous medium and the stress t_C at the particle contacts. Depending on $t_C - u_C$ curve and on the values of the elastic constants, the $\sigma_A - u_A$ curve may have either ascending or descending branches. Of special interest is the case when $t_C = p$, $\tau_B = -p$, and, thus, $\sigma_A = 0$: the element is subject to self-stressing. Depending on the value of the eigen stresses p, the $\sigma_A - u_A$ curve behaves differently: it is possible that $\Delta\sigma_A > 0$, $\Delta u_A > 0$, i.e., the curve first ascends and then descends. Under high p it can be that $\Delta u_A > 0$ while $\Delta\sigma_A < 0$. In this case, the element behaves as a source of energy.

We introduce micro-variables to describe averaged micro-strains and micro-stresses in grains, $\begin{pmatrix} \varepsilon_{11}^t \\ \varepsilon_{22}^t \\ \varepsilon_{12}^t \end{pmatrix} = T \cdot \begin{pmatrix} t_{11} \\ t_{22} \\ t_{12} \end{pmatrix}$, $\begin{pmatrix} \varepsilon_{11}^\tau \\ \varepsilon_{22}^\tau \\ \varepsilon_{12}^\tau \end{pmatrix} = T \cdot \begin{pmatrix} \tau_{11} \\ \tau_{22} \\ \tau_{12} \end{pmatrix}$, in porous medium, $\begin{pmatrix} \varepsilon_{11}^p \\ \varepsilon_{22}^p \\ \varepsilon_{12}^p \end{pmatrix} = P \cdot \begin{pmatrix} p_{11} \\ p_{22} \\ p_{12} \end{pmatrix}$, at the grain contacts, $\begin{pmatrix} \varepsilon_{11}^R \\ \varepsilon_{22}^R \\ \varepsilon_{12}^R \end{pmatrix} = R \cdot \begin{pmatrix} t_{11} \\ t_{22} \\ t_{12} \end{pmatrix}$ (Figure 3), where the third order square matrixes T, P, R depend, respectively, on elastic constants of grains and porous medium, and on moduli of plastic sliding at the contacts of grains. The inter-grain sliding curve is approximated by a piecewise-linear function as is shown in Figure 4 [9, 10]. The introduced values obey the consistency conditions given by

$$\begin{pmatrix} \overset{*}{\sigma}_{11} \\ \overset{*}{\sigma}_{22} \\ \overset{*}{\sigma}_{12} \end{pmatrix} = \begin{pmatrix} t_{11} \\ t_{22} \\ t_{12} \end{pmatrix} + 2 \cdot m \cdot \begin{pmatrix} p_{11} \\ p_{22} \\ p_{12} \end{pmatrix} = \begin{pmatrix} t_{11} \\ t_{22} \\ t_{12} \end{pmatrix} + 2 \cdot (1-m) \cdot \begin{pmatrix} \tau_{11} \\ \tau_{22} \\ \tau_{12} \end{pmatrix}, \quad (6)$$

$$\begin{pmatrix} \overset{*}{\varepsilon}_{11} \\ \overset{*}{\varepsilon}_{22} \\ \overset{*}{\varepsilon}_{12} \end{pmatrix} = \begin{pmatrix} \varepsilon_{11}^t \\ \varepsilon_{22}^t \\ \varepsilon_{12}^t \end{pmatrix} + \begin{pmatrix} \varepsilon_{11}^R \\ \varepsilon_{22}^R \\ \varepsilon_{12}^R \end{pmatrix} = (1-m) \cdot \begin{pmatrix} \varepsilon_{11}^\tau \\ \varepsilon_{22}^\tau \\ \varepsilon_{12}^\tau \end{pmatrix} + m \cdot \begin{pmatrix} \varepsilon_{11}^p \\ \varepsilon_{22}^p \\ \varepsilon_{12}^p \end{pmatrix}. \quad (7)$$

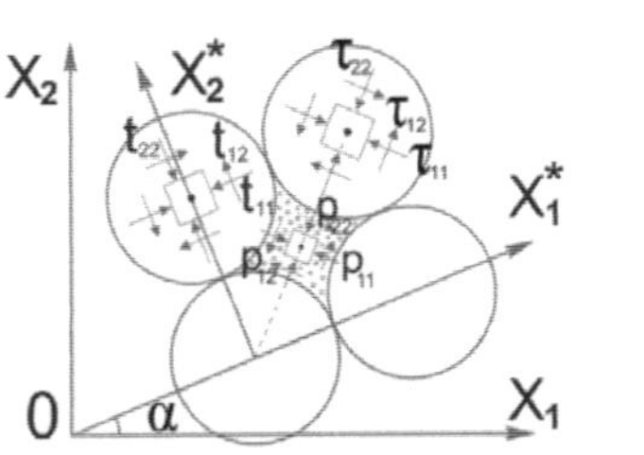

Figure 3. Internal structure of the medium (orientations of effective forces).

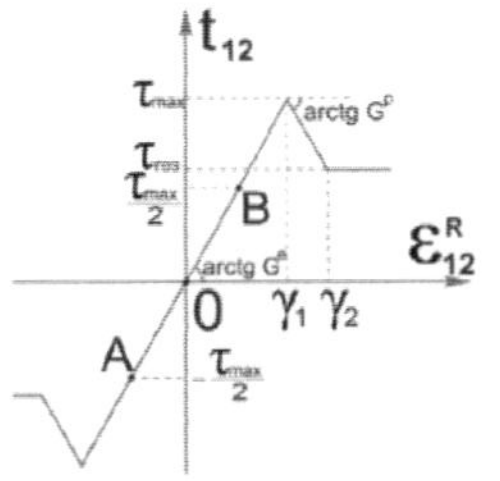

Figure 4. Inter-grain sliding curve.

Here, $\overset{*}{\varepsilon}_{ij},\overset{*}{\sigma}_{ij}$ are the components of the averaged macro-strains and macro-stresses in the local coordinates connected with the orientation of the grain matrix, $0<m<1$ is understood as a characteristic of clear opening. After elimination of the micro-variables, the constitutive relations will be given by

$$\begin{pmatrix}\varepsilon_{11}\\ \varepsilon_{22}\\ \varepsilon_{12}\end{pmatrix}=W\cdot\left((T+R)^{-1}+2\cdot(T+P)^{-1}\right)^{-1}\cdot W^{-1}\cdot\begin{pmatrix}\sigma_{11}\\ \sigma_{22}\\ \sigma_{12}\end{pmatrix}, \quad (8)$$

where $\varepsilon_{ij},\sigma_{ij}$ are the components of the averaged macro-strains and macro-stresses in arbitrary rectangular coordinates; the matrix W is reflective of the orientation of the grain matrix relative to the rectangular coordinates and depends on α (refer to Figure 3).

Based on the described approach and the mathematical model of a structurally nonuniform rock mass (8), the finite element algorithm and software program are developed to solve 2D boundary value problems, considering internal self-balancing stresses.

3. Numerical model

This section exemplifies calculation of a geomaterial compression. Let a rectangular specimen (see Figure 5) be compressed between two parallel plates displaced vertically. The boundary conditions are given by

$$\begin{aligned}&\Delta u_2\big|_{\Gamma_1^+}=\Delta d,\quad \Delta u_2\big|_{\Gamma_1^-}=-\Delta d,\\ &\Delta u_1\big|_{\Gamma_1^+}=\Delta u_1\big|_{\Gamma_1^-}=\Delta\sigma_{11}\big|_{\Gamma_2}=\Delta\sigma_{12}\big|_{\Gamma_2}=0,\end{aligned} \quad (9)$$

where Δu_i, $\Delta\sigma_{ij}$ are the increments in the displacements and stresses at the certain segment of the boundary; Δd is the increment of the loading parameter (due to nonlinearity, the problem is solved in the quasi-static formulation, by loading steps). Let the parameter of the orientation of the grain matrix be set as $\alpha=\pi/4$ and the parameter of the clear opening—as $m=0.5$. All stresses are referred to the value of a maximum shear stress at the inter-grain contacts, $\tau_{\max}$ (see Figure 4): $\tau_{\max}=1$, $\tau_{res}=0.4$. It is noteworthy that with such orientation of the grain structure, the influence of self-balancing shear stresses in the medium is the most pronounced.

The aim of the numerical modeling is to assess influence of the self-balancing (eigen) stresses. To this effect, three calculation series are performed for three different scenarios of setting the self-balancing stresses. In all series, the calculations end upon reaching one and the same value of the loading parameter, namely, $dH/H=-0.012$.

The first calculations series is performed for a geomaterial specimen without eigen self-balancing stresses, i.e. all components $t_{ij}=p_{ij}=0$ at the start time and, consequently, $\sigma_{ij}=0$. In this case, each point of the medium is in the position of the point O in the plot of the contact interaction of grains in Figure 4. Figure 5a illustrates the calculation results. Hereinafter, the unshaded areas show the state of strengthening at the contacts: $0<t_{12}<1$, $\varepsilon_{12}^R<\gamma_1$; the light-grey color marks the zones of weakening between grains: $0.4<t_{12}<1$, $\gamma_1<\varepsilon_{12}^R<\gamma_2$; the dark-grey color depicts the regions of the residual shear strength: $t_{12}=0.4$, $\varepsilon_{12}^R>\gamma_2$.

Let the start-time shearing be set as $t_{12}=-0.5$ at the contacts of grains and as $p_{12}=0.5$ in the porous medium. The other components of the micro-stress tensors of grains and pores are assumed zero. In the plot in Figure 4, the pre-set initial conditions conform with the point A. Apparently, the internal micro-stresses balance each other so that all macro-stress components $\sigma_{ij}=0$. The eigen stresses are set so that they "resist" compression. This situation is illustrated in Figure 5b. It is evident that the pre-set self-balancing stresses greatly decelerate growth of the plastic strain zones: these zones cover smaller domain as against the case of the absent self-balancing stresses under the same external force. Figure 5c shows the calculation results with the alternate self-balancing stresses $t_{12} = 0.5$, $p_{12} = -0.5$, which fall at the point *B* in the plot in Figure 4. In this case, these stresses "contribute" to compression. From the comparison of the deformation patterns obtained under equal external loading in Figures 5a–5c, the eigen stresses exert a considerable influence on evolution of plastic strain zones, and can either decelerate (Figure 5b) or accelerate (Figure 5c) expansion of plastic deformation and failure of specimens.

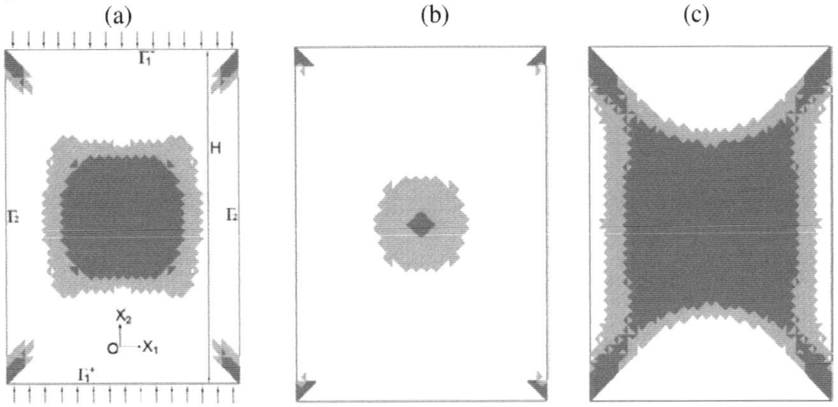

Figure 5. Plastic deformation zones: (a) without initial stresses included; (b) the initial stresses "resist" compression; (c) the initial stresses "contribute" to compression.

4. Conclusion
The proposed approach allows mathematical modeling of deformation of structurally nonuniform media, considering internal self-balancing stresses.

The internal self-balancing stresses have a considerable influence on the behavior of plastic deformation zones, and can both decelerate and accelerate failure.

5. Acknowledgements
The investigation was done under support of Russian Science Foundation, project № 16-17-10121

References
[1] Moroz AI 2004 *Self-Stress State of Rocks* Moscow: MGGU (in Russian)
[2] Birger IA 1963 *Residual Stresses* Moscow: Mashgiz (in Russian)
[3] *Organizational Standard STO 36554501-019-2009 Identification of Self-Stress State in Rocks NITs Stroitelstvo* Moscow (in Russian)
[4] Revuzhenko AF 2000 *Mechanics of Elastoplastic Media and Nonstandard Analysis* Novosibirsk: NGU (in Russian)
[5] Revuzhenko AF 2012 *Mathematical Analysis of Functions of Non-Archimedean Variable: Application-Specific Mathematical Tool for Description of Structural Levels in a Geomedium* Novosibirsk: Nauka (in Russian)
[6] Kurlenya MV, Oparin VN, Revuzhenko AF and Shemyakin EI 1987 Some feature of rock response to near-range blasting *Doklady AN SSSR* Vol 293 No 1
[7] Revuzhenko AF 1990 Rock as a medium with internal energy source and sinks. Report 1 *Journal of Mining Science* Vol 26 No 4 pp 301–308
[8] Trusov PV 2009 Some issues of the nonlinear mechanics of a deformable solid (Disputable) *Matemat. Model. Sistem Protsessov* 2009 No 17

[9] Lavrikov SV, Mikenina OA, Revuzhenko AF and Shemyakin EI 2008 Concept of non-Archimedean multiscale space and model of plastic media having structure *Fizich. Mezomekh.* Vol 11 No 3
[10] Lavrikov SV, Mikenina OA, Revuzhenko AF 2008 Rock mass deformation modeling using the non-Archimedean analysis *Journal of Mining Science* Vol 44 No 1 pp 1–14

Continuous monitoring of deformations in mines using laser interferometer–deformograph

MD Parushkin[1] and DO Tereshkin[2]

[1]Institute of Laser Physics, Siberian Branch, Russian Academy of Sciences, Novosibirsk, Russia
[2]Siberian Division, Geophysical Service, Siberian Branch, Russian Academy of Sciences, Novosibirsk, Russia

E-mail: [1]mihail.parushkin@yandex.ru; [2]denis@kasan.ws

Abstract. The article gives a detail description of the deformation graphing equipment designed at the Institute of Laser Physics, SB RAS, for the geodynamic monitoring of stress state in mine galleries and other underground openings, and for the analysis of ground surface vibrations before earthquakes and other seismic events. The more comprehensive study of separate rock mass blocks uses the advanced modification of the prototype deformograph introduced into service on Talgar testing ground in Kazakhstan. The programming support of the equipment is described.

1. Introduction

The Institute of Laser Physics, SB RAS, has developed and introduced a high-precision laser deformography equipment intended for geodynamic monitoring of stresses and strains in rocks in mines and ground vibrations before earthquakes and other seismic events [1–3]. The measurement equipment is based on the layout of the Michelson unequal-path interferometer with the shift of optical frequency range into radio frequency range using the method of optical heterodyne detection. The equipment performs continuous recording of time variation in the phase of a light wave due to Doppler effect when the light wave is reflected from a moving mirror (mounted on a test object) [4, 5]. The shift is determined as the difference of phases of two radio signals from light-detecting device in reference and measurement optical channels to electronic circuit of phase detector. A variation compensation unit for refracting index of air enables measurement in underground excavations without auxiliary vacuum or excess pressure beam waveguides at measurement error ~ $2X10^{-9}$ ($\Delta L/L$). The limit sensitiveness of shift of the equipment is $\lambda/4096$ μm (where $\lambda = 0.6328$ μm is the gauge length of laser). The deformograph is designed for operation in the mode of continuous recording and short-term shutdown for the routine maintenance at a frequency of once per 3 months. Hermetic enclosures of mechano-optical and electronic units ensure long-term operation of the equipment under conditions of 100% moisture content of air.

At the present time, two deformography equipment sets have been put into practice. The first set operates on the testing ground of the seismic station Talaya situated on the south-west side of the Baikal Lake (51°40′N, 103°38′E), the second set is run at the geophysical observatory Talgar in Kazakhstan (43°18′N, 77°14′E). The second, later version of the deformograph possesses higher sensitiveness and sampling frequency [6]. This paper describes performance and measurement procedures of the latter version of the deformography equipment.

The deformography equipment is a set of optical, mechanical and electronic units. Figure 1 shows the layout of a two-channel laser deformograph, including laser, laser power supply, interferometer, mechano-optical compensation block, automatic laser frequency and phase control, electronic unit of phase shift measurement and data transmission, mirrors at the ends of optical paths and mechano-optical block of wavelength standard. Also, the measurement equipment includes a software support for the recording system and utility programs for the data pre-processing. The crucial components of the system are described below in this paper.

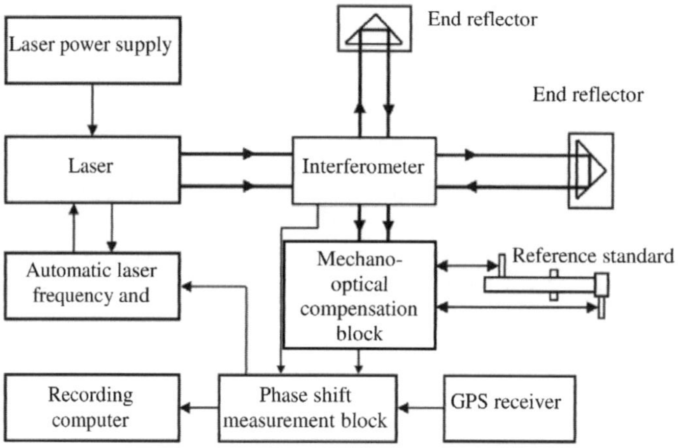

Figure 1. Layout of laser deformography equipment set.

2. Laser block

The laser source provides radiation for heterodyne detection and frequency measurements with the difference between optical frequencies of 1 MHz phase-locked with the frequency of a reference generator. The laser sources are represented by two He–Ne lasers with the wavelength of 0.6328 μm and capacity of 1 mW meant for continuous operation within 10–15 thousand hours. The phase synchronization and stabilization of the difference frequency between measurement and heterodyne lasers uses the system of frequency–phase control over piezoceramic elements, on which mirrors of laser cavities are mounted, by means of changing the optical lengths of the cavities and, thus, the laser radiation frequency. The structure of this block allows room for additional phase-locked He–Ne laser or laser frequency standard to stabilize operating wavelengths of basic lasers.

3. Interferometer block

The interferometer block changes phases of optical signals into electric signals proportional to the change in length of optical path of probing beams along measurement lines. The interferometer contains three independent optical channels, one of which is the reference channel, and the other two are the measurement channels. They are based on the layout of Michelson interferometer where the reference beam is the heterodyne laser radiation with the frequency shifted by 1 MHz relative to the measurement laser frequency. The reference optical channel receives electrical signal with the operating frequency of 1 MHz relative to which measurement of phases of information electrical signals from the photo-detectors of the measurement channels is executed. Figure 2 shows the optical circuit of the interferometer. Beam 1 means the measurement radiation, beam 2—heterodyne radiation. The reference signal is sent from photo-detector PD3, the measurement signals are from photo-detectors PD1 and PD2.

The block includes passive temperature drift compensation, which reduces instrumentation error of the measurement circuit under outward temperature variation down to < 0.01 μm/deg. The heterodyne radiation is sent to mirrors 2 and 3. The radiation beam passes optical paths between errors 1–2, 2–3 and 3–4. In such sequence, thermal distortions of sections of metal plate where optical elements 1, 2 and 3 are placed (and, accordingly, the change in the optical path length) are equal to thermal distortions of sections of the plate between optical elements 5, 6 and 7 and the contact zone of the plate and the measurement block legs fastened to the plate.

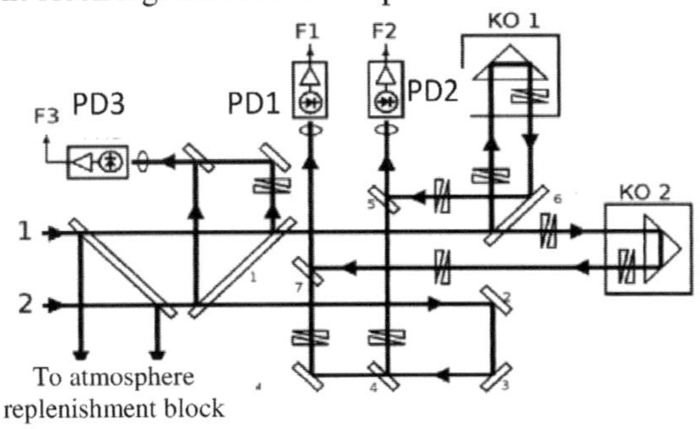

Figure 2. Optical circuit of the two channel interferometer. Description is given in the text.

The block can also use optical fiber waveguides to feed measurement radiation to the measurement optical circuit. The fiber optics simplifies operation of the equipment and eliminates tuning of the optical circuit after replacement of active laser elements and rigid fastening of the laser source on the plate.

4. Mechano-optical compensation block

The block of mechano-optical compensation of air influence measures relative fluctuations in the refractive index of air in the course of deformation measurements. The sensing device in the compensation block is designed as the heat-compensated length standard, which enables measurement without frequency-stabilized laser sources.

Such design enables elimination of air influence and compensation of measurement error caused by the change in the measurement wave length of the probing laser radiation. The sensing device is manufactured as a self-sustained unit and can be arranged at a distance to 10 m from the interferometer.

The compensation block circuit records the change in the length of optical path equal to the spacing of mirrors mounted on the length standard. The geometrical distance between the mirrors only changed by the value of thermal distortion of the material the length standard is made of, and is immune to the influence of deformations of a test object.

The basic circuit of the mechano-optical compensation block is shown in Figure 3. The block is composed of interferometer and thermally compensated length standard. The interferometer transform optical signal into electrical signal with the value proportional to the change in the refractive index of air and to the variation in the laser radiation wave length. This value is introduced automatically as a deformation measurement correction. The interferometer has two measurement channels based on the layout of the Michelson unequal-path interferometer with the transmission of phase data from the optical range to the radio frequency range using the method of optical heterodyning.

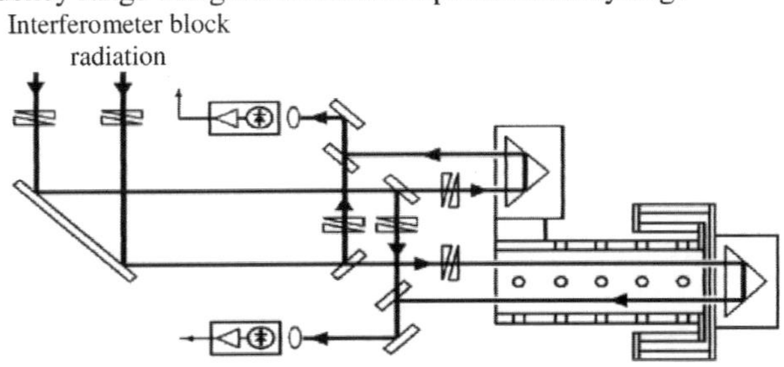

Figure 3. Basic optical circuit of the mechano-optical compensation block.

The length standard 1 m long is made of 36 N alloy thermally treated in a special mode and with a thermal expansion coefficient of 8×10^{-7} 1/deg, and is equipped with an auxiliary passive thermal compensation unit made of aluminum alloy. Such design ensures constant distance between the reflectors with an accuracy of 1×10^{-9} ($\Delta L/L$) under the variation in air temperature in the range of ±10 deg.

5. End reflectors

The end reflectors of the laser deformograph (ER1 and ER2 in Figure 2) ensure sustainable measurements at essential angular and linear displacements of a test object (plate) at which they are mounted. The reflectors include prisms with additional alignment units on optical wedges, which allow considerable relaxation of prism manufacture accuracy standards. The operating light aperture of the reflectors is 25 mm.

6. Phase meter

The phase meter is intended for primary transformation of two input electrical signals with the frequency of 1 MHz fed from the channels of the optical circuit into a digital signal with the value proportional to the phase shift between these signals. The deformograph uses the 24-bit 6-channel phase meter with the measurement accuracy of $\pi/1024$ at the frequency of 40 kHz and $\pi/8192$ at the frequency of 5 kHz.

7. Software support

The software support of the laser deformograph is composed of two OC WINDOWS program packages for recording and pre-processing of data.

The data acquisition system includes phase meter, seismic signal recorder, data collection server, data visualization program, configuration manager and display. The data acquisition system receives data from the measurement systems, performs decoding, storage and visualization of the data, and detects seismic events based on LTA/STA criterion.

The data pre-processing program performs editing and on-line analysis of the deformograph data aimed to detect anomalous deformation processes. The analysis embraces digital arrays of data on variation in phase signals obtained from the two independent measurement channels and one compensation channel of the deformograph. Deformations in the two independent orthogonal directions are detected by means of program-aided diminution of the phase signals recorded in a digital form in the measurement channels and the compensation arm, with an adjustment coefficient proportional to the ratio of geometrical lengths of the measurement and the compensation arms. Moreover, the analysis includes a time series of the difference deformation between the first and second measurement arms, obtained by direct deduction of the phase signals of the phase meter.

The onward detailed processing of the resultant series uses the dedicated utility programs.

8. Acknowledgements

The investigation was done under support of Russian Science Foundation, project № 16-17-10121

References

[1] Bagaev SN, Orlov VA, Fomin YuN and Chebotaev VP 1992 Heterodyne laser deformographs for high-precision measurements *Fizika Zemli* No 1

[2] Orlov VA, Panov SV, Parushkin MD, Semibalamut VM and Fomin YuN 2013 High-sensitive laser measurement capacities in the analysis of seismic and geodynamic events *Geodynamics and Stress State of the Earth's Interior: Conf. Proc* Novosibirsk: IGD SO RAN (in Russian)

[3] Panov SV, Parushkin MD, Rybushkin AYu Semibalamut VM and Fomin YuN 2015 Outdoor laser measurement of strains *Fund. Prikl. Vopr. Gorn. Nauk* 2015 No 2 pp 332–338

[4] Bagaev SN, Dychkov AS, Om AE, Orlov VA, Fomin YuN and Chebotaev VP *RF Patent No. 362923 Dual-Frequency Interferometric System to Measure Linear Displacements*

[5] Bagaev SN, Orlov VA, Rybushkin AYu, Semibalamut VM amd Fomin YuN *RF Patent No. 2082085 Dual-Frequency Interferometric Measurement System for Linear Displacements*
[6] Orlov VA, Semibalamut VM, Panov SV, Parushkin MD, Tereshkin DO et al 2013 Date of the first laser deformograph observations in Talgar gallery on Alma-Ata prognostic test ground *Vestnik KazNAEN* No 3 pp 79–83

Features of stress state of support under varying displacement of free contour of underground excavation

VM Seryakov

Chinakal Institute of Mining, Siberian Branch, Russian Academy of Sciences, Novosibirsk, Russia

E-mail: vser@misd.nsc.ru

Abstract. The scope of the discussion covers the issues of stress state assessment in support and rock mass surrounding an underground mined-out void, considering partial relaxation of rocks from initial stresses prior to contact interaction between the underground excavation contour and the support elements. The dedicated methods and algorithms are used to determine stress field distribution in elements of support in an arched underground excavation under varying pre-contact displacements of the excavation contour.

Stability assessment in mining with reinforcement and support includes estimation of stresses both in surrounding rock mass and in support units [1]. These issues are sufficiently studied, with formulation of boundary value problems on mechanical condition of rocks and support, and with analytical calculations and numerical procedures meant to determine behavior and features of stresses in rocks and support under varied loading conditions [1–5]. On the other hand, some assumptions on rocks–support system deformation not to the full describe conditions in a real mine and need refinement in view of increasingly complicating ground conditions in deeper level mining [5].

It is important that a rock–support interaction model correctly describes mechanics of this contact interaction [1–5]. One of the common formulations of the problem on stress state of surrounding rock mass and support units considers deformation of discontinuous rock mass at rigid contact with support when the back surface of the support is subjected to the action of forces equivalent to in situ stresses. Mechanically, such formulation means that support is installed "instantaneously" and no elastic unloading of rock mass and in situ stress relief takes place [6, 7]. Technically, "instantaneous" installation of mine support immediately after drivage is infeasible. There is always a time gap between drivage and support installation, and surrounding rock mass undergoes displacement and unloading of in situ stresses during this time.

Depending on support installation technology, destressing of surrounding rocks can be partial or complete. Under partial unloading, exposed surface displaces a little inwards the mined-out void and surrounding rocks undergo unloading of elastic stresses. After support installation, stress relief continues in rocks, but in this case, both surrounding rocks and support units are subjected to deformation. Apparently, if surrounding rocks less displace inwards an underground excavation, support units will be subjected to lower loading later on. Applied problems use simpler formulations, as a rule. First, a problem on deformation of surrounding rock mass and unsupported excavation is solved. Then, support with its back surface subjected to the action of normal compressive stresses is considered. Solutions of the two problems with an additional condition of equal displacement of the back surface of the support and the exposed surface of surrounding rock mass are assumed a final solution [2, 3]. It is difficult to assess the degree of approximation due to consideration of unrelated deformation of surrounding rocks and support units.

Some formulations, accounting for support installation after partial relief of in situ stresses in surrounding rocks, determine displacement of back surface of support using relations: $u–u_0 = u_p$; $v–v_0 = v_p$, where u, v—projections of displacement vectors of unsupported exposure in the Ox and Oy axes (in 2D problem); u_0, v_0—projections of displacement vectors of exposed surface before support installation; u_p, v_p—displacements of back surface of support after installation [8]. In such formulation, displacement of surrounding rocks in combination with deformation of support has the same values as in case of an unsupported excavation. In other words, support has no influence on total displacement of surrounding rocks. In order to assess the effect of such approximation on the results of stress calculation in surrounding rocks and support units, it is required to formulate problems that account for rock–support contact interaction.

The Institute of Mining, SB RAS, offers developed methods and algorithms to model stress distribution in rocks and support, considering sequence of mining operations [9, 10]. The algorithms use initial stress and initial strain methods with a stiffness matrix that is unchanged in solving nonlinear problems of deformable solid mechanics [11, 12]. Thereupon, a method to calculate stresses and strains in support units and surrounding rock mass with regard to free displacement of rocks until the contact with back surface of support has been proposed and approved. Modeling the process of formation of an underground excavation using the initial stress method and equaling all stress components in finite elements representing the mined-out void zero enables splitting the problem into two sub-problems. The first sub-problem determines mechanical condition of rocks in the course of free deformation and displacement inwards the excavation; the problem is assumed solved when displacements reach a certain value governed by the support installation technology. The second sub-problem describes joint deformation of surrounding rocks and support using the initial stress method, as well.

The applicability of the method to calculation of stresses and strains in surrounding rocks and support with regard to support installation conditions was illustrated in [9, 10], with the assessment of stress behavior in surrounding rocks depending on the value of free displacement of exposed surface inwards a mine excavation until the contact with the support, and with the discussion of the method application when rocks and support posses essentially different physical properties.

This article presents the results of using the mentioned algorithms to estimate stresses in arch support in case of different free displacements of surrounding rocks until the contact with the support. Figure 1 shows the configuration of underground excavation and support. The calculations are performed for plane strain conditions. At the vertical boundaries of the computational domain, the horizontal displacement vector component u and the shear stress tensor component τ_{xy} are set zero. Such boundary conditions conform with the in situ stress state of rock mass with the stress tensor composed of $\sigma_y^0 = -\wp H$; $\sigma_x^0 = -v \wp H/(1-v)$; $\tau_{xy}^0 = 0$ and are observed in the regions where there is no tectonics [11]. Here, σ_x^0, σ_y^0, τ_{xy}^0—normal and shear components of the stress tensor; $\wp$—bulk density of rocks; v—Poisson's ratio of rocks; H—depth below the ground surface. The Ox axis is oriented horizontally and Oy—vertically. The upper boundary of the computational domain is free from external load. At the lower boundary, the vertical displacement vector component v and the shear stress tensor component τ_{xy} are assumed zero. The physical properties of rock mass are: $E = 50000$ MPa; $v = 0.25$. For support, it is assumed that $E = 150000$ MPa; $v = 0.25$. The bulk density of rocks and support is 0.03 MN/m^3.

Figure 1 shows the isolines of the principal stresses σ_1 and σ_2 in support and in a small area of rock mass surrounding an arch excavation in the case when rocks are not unloaded from in situ stresses before support installation. Compression concentration zones of σ_2 are observed in support at the sidewall bottom, starting from the contact with the excavation floor. The highest compressive stresses are observed on the back surface of the support. In the arch of the excavation, σ_2 are lower. The stresses σ_1 are not higher than –5 MPa in modulus in the support and are lower than –30 MPa in the concentration zones (in modulus). In the arch of the excavation, small areas of tensile stresses not higher than 1 MPa appear in the support. This stress pattern is in the whole in conformity with the known mechanical behavior of support installed in arch excavations [2, 3].

(a) (b)

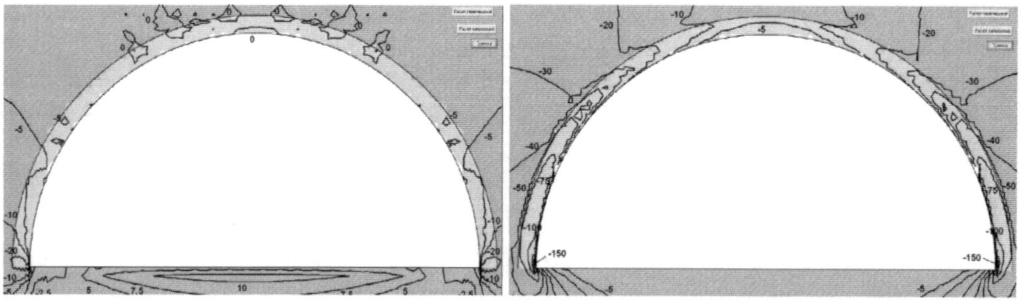

Figure 1. Principal stresses (a) σ_1 and (b) σ_2 (MPa) in support in case of its "instantaneous" installation.

Qualitatively, the stress pattern in the support in case of the partial unloading of rocks during free displacement inwards the mined-out void until the contact with the support remains the same (see Figure 2). Displacement of surrounding rocks in this case is half the displacement of surrounding rocks without support. The concentration zones of the compressive stresses σ_1 and σ_2 diminish as the compression reduces. The concentration zone of σ_2 shifts to the back surface of the support and the stresses here are not higher than 100 MPa (in modulus). Almost entire support is unloaded from the compressive stresses σ_1 not higher than –5 MPa (in modulus).

(a) (b)

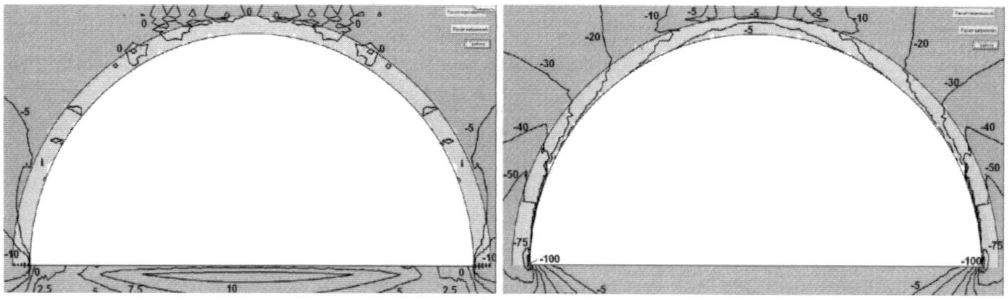

Figure 2. Principal stresses (a) σ_1 and (b) σ_2 (MPa) in support in case of surrounding rock displacement making the half of displacement of unsupported exposure.

This change in the stress pattern in the support even deepens with an increase in stress relief in surrounding rocks before the support installation. Figure 3 depicts stress patternt in support in case of 95% displacement of surrounding rock mass until its contact with the support, which is conformable with deformation of unsupported excavation. In this case, the support is stress-free. At the same time, there is a small and low-stress concentration zone of σ_2 at the junction of the support and the excavation floor. Thus, the compression concentration zone in an arch excavation with support at different displacements of surrounding rock mass until its joint deformation with the support is at the floor of the excavation.

(a) (b)

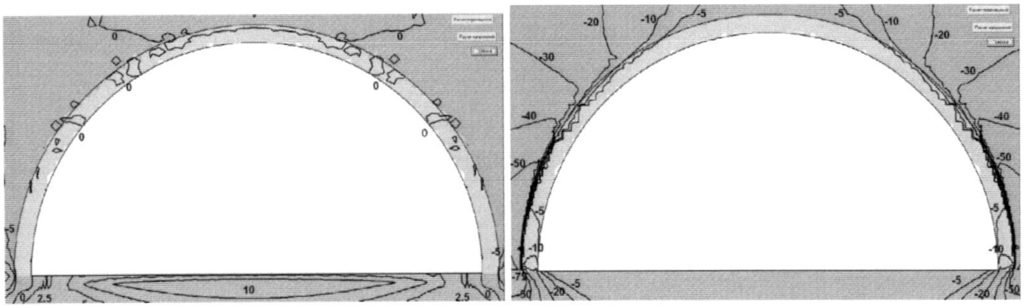

Figure 3. Principal stresses (a) σ_1 and (b) σ_2 (MPa) in support in case of surrounding rock displacements equaling 95% displacement of rocks in unsupported excavation.

Conclusion

The discussed method to assess stress state in the rock mass–support system under deformation makes it possible to determine loading on the support and to analyze its stress state in case of different displacements of adjoining rock mass.

Under gravitational in situ stresses, the highest compression zone in a support installed in an arch excavation is at the juncture of the sidewalls and the floor of the excavation. The concentration of the compressive stresses in this zone is preserved under further increase in displacements of surrounding rock mass.

References

[1] Baklashiv IV and Kartozia BA 1984 *Mechanics of Underground Structures and Support Units* Moscow: Nedra (in Russian)
[2] Bulychev NS 1989 *Mechanics of Underground Structures in Problems and Examples* Moscow: Nedra
[3] Bulychev NS 1994 *Mechanics of Underground Structures* Moscow: Nedra (in Russian)
[4] Kartozia BA Fedunets BI, Shuplik MN et al 2003 *Mine and Underground Construction: University Textbook* Moscow: Gornaya Kniga (in Russian)
[5] Protosenya AG, Dolgy IE and Ogorodnikov YuN 2003 *Mine and Underground Construction in Problems and Examples* Saint-Petersburg: Plekhanov's Mining Institute (in Russian)
[6] Nasonov ID, Fedyukin VA and Shuplik MN 1992 *Underground Construction Technology* Moscow: Nedra (in Russian)
[7] Turchaninov IA, Iofis MA, and Kasparyan EV 1989 *Rock Mechanics* Leningrad: Nedra (in Russian)
[8] Protosenya AG, Ogorodnikov YuN, Demenko PA et al 2011 *Mechanics of Underground Structures* Saint-Petersburg: SPPGU-MANEB (in Russian)
[9] Seryakov VM 2008 Implementation of the calculation method for stress state in rock mass with backfill *Journal of Mining Science* Vol 44 No 5 pp 439–450
[10] Seryakov VM 2015 Calculating stresses in support and sidewall rocks in stagewise face drivage in long excavations *Journal of Mining Science* Vol 51 No 4 pp 673–678
[11] Zienkiewicz O 1971 *The Finite Element Method in Engineering Science* London: McGraw Hill
[12] Fadeev AB 1987 *The Finite Element Method in Geomechanics* Moscow: Nedra (in Russian)

Rock mass diagnostics based on microseismic monitoring data at Sheregesh deposit

TV Lobanova[1], GL Lindin[2], OL Trofimova[1], SK Shultaev[3] and VV Prib[3]

[1]Siberian State Industrial University, Novokuznetsk, Russia
[2]Novokuznetsk Institute–Division, Kemerovo State University, Novokuznetsk, Russia
[3]Gornaya Shoria Division, EVRAZRUDA, Sheregesh, Kemerovo Region, Russia

E-mail: [1]lobanova_tv@sibsiu.ru

Abstract. Distribution of seismic events in rocks mass around a mineral deposit is studied, it is found that the distribution features conform with the local blasting operations and geology. Epicenters of concentration zones of seismic events are detected in the area of stoping and faulting between two ore locuses. Weakening planes generated under four chronologically ordered shocks are indentified. Occurrence of these planes is characterized, and the stress redistribution after blasting is analyzed. The authors apply an integrated approach to rockburst hazard prediction, including calculus of apparent velocity of migration of seismic events during the first ten minutes after a blast and the rockburst probability assessment.

Sheregesh iron ore mining with underground method is carried out in difficult ground conditions, with exfoliation, balmstones (overhanging ledges), spalling, shocks and micro rock bursts. Major dynamic events due to rock pressure take place in the influence zone of stoping, nearby large faults and blasting sites in production ore blocks. This is for the first turn associated with the concentration of stresses in stopes, gangue pillars and enclosing rock mass. The phenomenon is aggravated under seismic effect of blasting when nonuniform loading of faults and other areas of structural weakening results in displacement of rocks along block boundaries and initiates dynamic events.

Sheregesh iron ore field is rockburst-hazardous [1]. In accord with the Federal Norms and Rules of Industrial Safety *Regulations on Safe Mining at Rockburst-Hazardous Deposits* [2] and the related *Guidelines* [1], general assessment of the deposit uses the regional prediction methods, and rockburst hazard estimation in specific areas involves the local prediction techniques. Microseismicity observation station in Sheregesh Mine of the Gornaya Shoria Division of EVRAZRUDA is one of the regional methods of stress–strain state prediction in Sheregesh ore field. On the other hand, that seismic station only performs recording of seismic events, with the definition of date, time, location and energy. Stress state prediction is neglected.

In recent years, the Siberian State Industrial University and the Novokuznetsk Institute–Division of the Kemerovo State University jointly with Sheregesh and Tashtagol Mines accomplish research work aimed to develop procedure for assessment and prediction of rockburst hazard based on the data of microseismicity monitoring [3–6]. This paper gives generalized features of seismic events occurred in New Sheregesh and Understream orebodies of Sheregesh Mine in 2014–2016, based on which rockburst hazard of these areas in the course of mining is assessed and predicted. The urgency of the prediction is caused by the increase in number of bumps and low-frequency seismic events in the

Understream orebody late in 2015–early in 2016 as compared with 2014: the effect of these events has been felt underground and by the residents of Sheregesh settlement on the ground surface. That situation complicates mining and heightens social strain.

New Sheregesh and Understream orebodies are under high-rate mining: 90% of stoping operations are carried out there. Furthermore, the orebodies are developed using concurrently two mining methods: room-and-pillar and sublevel caving with self-propelled machinery. In the long view, it is planned to change completely to the sublevel caving: major ore reserves will be extracted with this mining method in 2016.

In 2014–2016 stress state of the local ore field was generated under the influence of large-scale production blasting (LPB) in blocks 7–9 in Understream orebody and blocks 300, 24 and 25 in New Sheregesh orebody. Preparation of ore blocks for caving used process blasting (PB). Sublevel caving operation in these orebodies also used PB.

The analyses of seismic events in the Sheregesh Mine field in 2014–2016 revealed distribution of the events in conformity with the blasting operations and geology of the area.

A geological feature of Sheregesh ore field is its enclosing rock mass that is composed of rocks (skarn, diorite, porphyrite and other) that have strength and elasticity higher than ore by 1.5 times, though the ore is also rather strong. This geological feature is of concern in the assessment of conditions and behavior of probable dynamic events due to rock pressure. Another difficulty is presented by blind occurrence of New Sheregesh and Understream orebodies.

Structural geology is also of importance. Exploration and geodynamic zoning detected a fault between New Sheregesh and Understream orebodies. The fault is located based on traces of fill of a thick apophyse of syenites, along the contacts and in the middle of which zones of crush strike north-eastwards (50–60°) with a steep north-westward dip (670–85°). The fault zone is 15 m wide, the zone of jointing (influence zone) has a width of 60–140 m. The dislocation is an oblique-slip fault caused by tension and compression. Along this zone, Understream orebody is displaced north-eastward by 45–55°. The fault zone features violent jointing, the high-angle cleavage cracks have north-eastward and north-westward orientations.

The assessment of rockburst hazard in Sheregesh ore field used three criteria: distribution of epicenters of seismic events in rocks, apparent velocity of seismic events within the first 10 minutes after a PB and rockburst probability.

The criterion of distribution of epicenters of seismic events allows identifying rockburst-safe and rockburst-hazardous ore blocks. The former occur in the zone of concentration of seismic event epicenters (CSE), and the latter lie in the zones of structural blocks and potential energy concentration.

The rockburst hazard criterion based on apparent velocity of seismic events is the velocity 10^3 mm/s [7]. When the maximum apparent velocity of seismic events immediately after PB in an ore block exceeds 10^3 mm/s, the surrounding rock mass shows dynamic response to the blast. In case that this velocity noticeably grows on approach of LPB, rock mass releases some potential energy and rockburst hazard gradually lowers.

A rockburst probability p—the main index of rockburst hazard—is given by [5]

$$p = (N - n) / N \,,$$

where N is the number of seismic event recorded in a 500-m zone around an ore block up to LPB, starting from blocks extracted earlier or concurrently with this block in the same area; n is the number of seismic event recorded in a 100-m zone around the block during PB on the current level and below.

When p is under 0.9, the ore block and surrounding rock mass are rockburst-unhazardous.

Based on the distribution of epicenters of seismic events (see Figure 1), the authors defined the activity of the fault between New Sheregesh and Understream orebodies (dash-and-dot line in Figure), where CSE zone was observed in 20142016. The distribution of seismic events characterizes dynamic formation of weakened planes and shearing [8].

Figure 2 shows a weakened plane governed by two angles [9]: a dip angle δ between the fault plane and the horizontal plane (x, y) and a strike angle θ set off clockwise from the x-axis (oriented toward the North) until the extended line of intersection of the weakened and horizontal planes. This criterion allows finding the direction of flow and accumulation of tectonic energy.

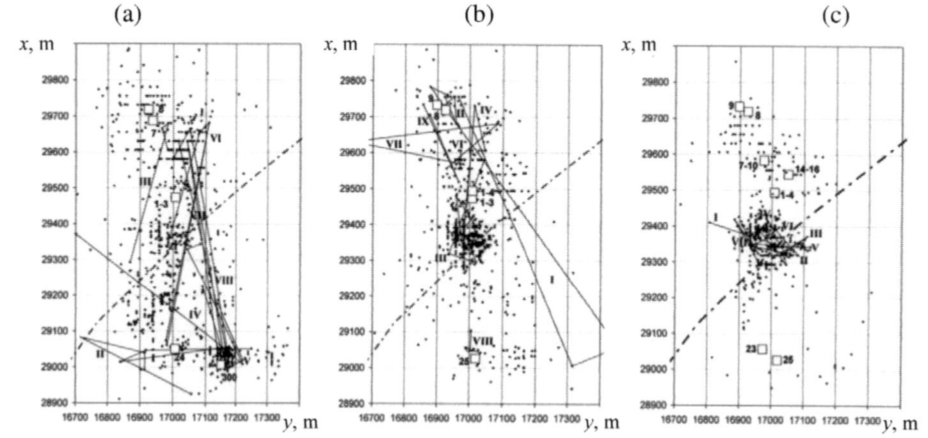

Figure 1. Distribution of seismic events in Understream orebody (above the dash-and-dot line) and New Sheregesh orebody (under the dash-and-dot line) in (a) 2014, (b) 2105 and (c) February 2016 with traces of weakened planes (solid lines): points—epicenters of seismic events; squares—centers of blocks under mining; dash-and-dot line—fault; Arabic figures—numbers of blocks; Roman figures—numbers of weakened planes.

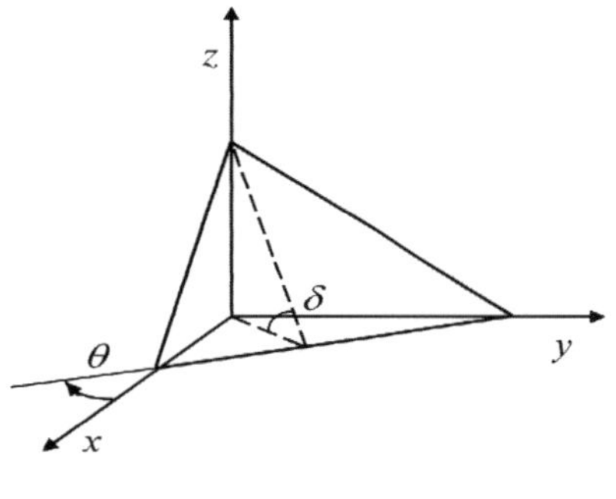

Figure 2. Elements of occurrence of a weakened plane.

Figure 1 shows the traces of the weakened planes in the form of seismic events and lines chronologically connecting them. In New Sheregesh and Understream orebodies, 8 and 9 weakened planes appeared in 2014 and 2015, respectively (see Table 1); these planes are activated by the process and production blasting.

It is noteworthy that CSE zone was generated near block 300 in 2014 (Figure 1a), and in 2015 and 2016 such zones were initiated in the area of the fault (Figures 1b and 1c). Rock mass in CSE zones contains many weakened planes and is incapable to accumulate potential energy. The weakened planes generated by mining in blocks 300, 1–3–120 and 1–4–164 in 2014 were reflective of stress redistribution between New Sheregesh and Understream orebodies and the fault. Planes I, V and VII were the channels for the tectonic energy to flow from CSE zone to periphery of New Sheregesh and Understream orebodies. Planes II, II and IV were the channels for the energy to flow from the fault inside New Sheregesh (planes II and IV) and Understream orebody (plane III). The longest planes appeared between the orebodies after LPB in block 300 and PB in blocks 1–4–20 and 1–3–164.

Table 1. Conditions and parameters of weakened planes in Sheregesh and Understream orebodies in 2014–2015.

Plane no.	Date (dd.mm.yy)	Time (h:min:s)	Operation	Dip, deg	Strike deg
			2014		
I	01.02.2014	20:03:53	31.01.2014—PB, explosive 9.0 t, block 300	49	106
II	04.02.2014	12:03:04		8	118
III	11.02.2014	17:33:18		50	167
IV	11.05.2014	16:22:18	26.04.2014—PB, explosive 8.6 t, block 300	7	106
V	23.05.2014	16:52:17	21.05.2014—PB, explosive 0.8 t, block 50	28	173
VI	25.10.2014	11:49:15	25.10.2014—PB, explosive 7/2 t, blocks 1–4–120	58	172
VII	07.12.2014	21:23:31	07.12.2014—LPB, explosive 257.8 t, block 300	71	10
VIII	29.12.2014	20:24:22	28.12.2014—PB, explosive 12.0 t, block 1–3–164	29	27
			2015		
I	18.01.2015	09:03:57	18.01.2015—LPB, explosive 139.0 t, block 8	15	24
II	18.01.2015	09:14:45		50	176
III	20.01.2015	10:48:02		25	77
IV	27.01.2015	01:54:40	25.01.2015—PB, explosive 2.1 t, block 1–3–164,	67	176
V	28.01.2015	10:07:16	PB, explosive 12.2 t, block 7	6	115
VI	12.06.2015	09:40:32	12.06.2015—LPB explosive 110.0 t, block 8	79	14
VII	13.06.2015	11:22:34		22	62
VIII	16.08.2015	08:06:59	16.08.2015—LPB, explosive 61.1 t, block 25	26	65
IX	12.09.2015	08:24:45	12.09.2015—LPB, explosive 52.0 t, block 9	58	172

In 2015 during mining in block 87, five weakened planes were found to intersect the block and CSE zone in enclosing rock mass. The planes had mostly north-westward and northward strike. The longest plane ran across both New Sheregesh and Understream orebodies. The other planes were mostly located in Understream orebody, between the fault and the actual stoping zone. These planes were the tectonic energy flow channels from CSE zone to the periphery of Understream orebody. It is noteworthy that major energy flew to the north-west of the orebody, in blocks 7, 8 and 9. That was probably connected with the tectonic stress field assessed by VostNIGRI [10]: $\sigma_1 = 2.6\sigma_3$; $\sigma_2 = 1.4\sigma_3$; $\sigma_3 = \gamma H$. The azimuth of the major stress σ_1 outside the influence zone of stoping, $A\sigma_1 = 335° \pm 15°$, was oriented northwestward in line of the orebody strike.

To correlate seismic events in 2014–2016 and blasting, Figure 3 shows the number of low-frequency events and bumps as against explosive weight in LPB and PB. It is clearly seen in the figure that the majority of bumps and low-frequency events took place after PB in block 8 (stages 1 and 2) and in block 9 (stages 3 and 4) on Level (+185) – (+255) m and after PB in blocks 4–1–164 and 7–10–164 on Level (+115) – (+185) m. The maximum number of bumps (184) and low-frequency events (1124) was recorded after PB in block 7–10–164 on February 25, 2016. The increase in the number of low-frequency events, especially in February 2016, is indicative of formation of a natural equilibrium arch in rock mass above the mined-out cavity in Understream orebody, which weakens the overlying strata and can induce movement in them up to the ground surface.

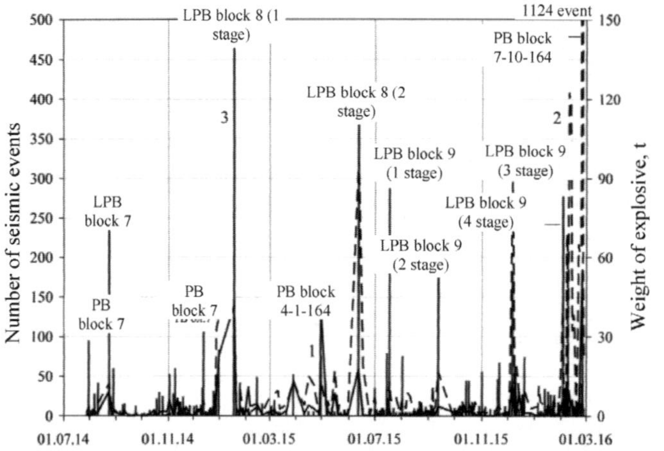

Figure 3. Time history of seismic events after blasting in Understream orebody: 1—bumps; 2—low-frequency seismic events; 3—quantity of explosive.

High activity of rock mass can result from simultaneous different-scale blasting in a confined environment of blind orebodies with large mined-out voids. Stress state of rocks mass around Understream orebody in 2015–2016 was generated by 6 large-scale blasts (2 LPB in block 8 and 4 LPB in block9).

The increased rockburst hazard near blocks 7, 8 and in Understream orebody was observed since 2014, after large-scale blasting in block 7 when rockburst probability exceeded the allowable value of 0.9 and equaled 0.935. Also in 2014, the recorded apparent velocity of seismic events was 162 mm/s at the criteria value of 1000 mm/s. The maximum rockburst probability was assessed after two stages of LP in block 9—0.937. By the end of 2015, this value reduced to 0.933. After the third-stage LPB in block 9 (December 6, 2015), the maximum apparent velocity of seismic events considerably grew from 224 mm/s (during 9 months of 2015) to 22909 mm/s (at the end of 2015). So, surrounding rock mass of block 9 dynamically responded to LPB and released accumulated potential energy, and rockburst hazard began lowering.

In the area of mining with a new technology and self-propelled machinery in Understream orebody, the rockburst probability was 0.871 late in 2015, which was lower than 0.9 but higher than in the course of previous 9 months of 2015 by 0.025. Although the insignificant increase in the rockburst hazard probability and the absence of seismic events within 10 min after PB, the situation meant higher rockburst hazard in the vicinity of sublevel caving area in the orebody.

On that background, LPB with 83 t of explosive was performed in block 9 (stage 4) on February 3, 2016 and PB with 12 t of explosive was carried out in block 7–10–164 on February 25, 2016, which critically increased the number of low-frequency seismic events and bumps in the enclosing rock mass. The bumps after that PB and until the next day PB in block 5–8–142 took place mostly in the influence zone of the fault between New Sheregesh and Understream orebodies. Seismic energy of the blasts activated numerous dislocations (weakened planes) governing the fault (see Figure 1c). Shearing started on those planes and induced various energy bumps. The planes were steeply dipping and shears occurred one after the other (Table 2). Apparently, shear along one plane activated shear along the next plane as in "domino", and the trigger was seismic energy of PB in block 7–10–164 200 m northward of the fault. The similar situation took place after each PB, while tens PB were executed each month plus 6 LPB in 2015–2016. Mostly bumps located in CSE zone around the fault that was a sink of energy of all LP and PB in New Sheregesh and Understream orebodies.

Thus, based on the stress assessment in rock mass around New Sheregesh and Understream orebodies, rockburst-hazardous zones were identified, which enabled resuming mining operations with the social tension removed.

Table 2. Weakened planes around the fault between New Sheregesh and Understream orebodies.

Weakened plane	Date (dd.mm.yy)	Time (h:min:s)	Bump energy class	Dip, deg	Strike, deg
I	25.02.2016	07:48:28	3.6	31	58
II	25.02.2016	08:42:41	3.6	4	51
III	25.02.2016	09:43:34	1.8	50	119
IV	25.02.2016	12:22:32	2.2	87	140
V	25.02.2016	12:50:27	2.7	77	87
VI	25.02.2016	15:09:07	1.3	65	99
VII	25.02.2016	16:12:10	1.7	79	40

Conclusion

It has been found that distribution of seismic events and generation of weakened planes activated by process and production blasting assists in determination of flow and accumulation of tectonic energy.

The causes of activation of seismic events (bumps and low-frequency events) after blasting are revealed.

The authors have illustrated feasibility of rockburst hazard assessment and prediction based on integrated application of three rockburst hazard criteria: distribution of epicenters of seismic events, apparent velocity of seismic events during the first 10 min after a process blast and rockburst probability.

References

[1] Eremenko AA, Eremenko VA, Klimko VK, Lazarevich TI, LObanova TV et al 2015 *Gudielines on Safe Mining of Rockburst-Hazardous Iron Ore Bodies in Sheregesh* Novosibirsk–Novokuznetsk (in Russian)

[2] Federal Norms and Rules of Industrial Safety *Regulations on Safe Mining at Rockburst-Hazardous Deposits* approved by the Federal Environmental, Industrial and Nuclear Supervision Service of Russia Decree No 576 as of Dec 2 2013

[3] Lindin GL, Lobanova TI and Klimko VK 2008 Characteristics of sesimic events at Tashtagol iron ore deposit in 2005–2006 *Geodynamics and Stress State of the Earth's Interior: Conf. Proc.* Novosibirsk: IGD SO RAN (in Russian)

[4] Lindin GL and Lobanova TV 2010 Effect of weakened planes on rockbursting in Tashtagol Mine *Geodynamics and Stress State of the Earth's Interior: Conf. Proc.* Novosibirsk: IGD SO RAN (in Russian)

[5] Lindin GL and Lobanova TV 2012 Pre-rockbursting seismic activity in Tashtagol Mine *J. Min. Sci.* Vol 48 No 2 pp 260–268

[6] Lindin GL and Lobanova TV 2013 Energy sources of rockbursts *J. Min. Sci.* Vol 49 No 1 pp 36–43

[7] Oparin VN, Vostrikov VI, Tapsiev AP et al 2006 Kinematic criterion for prediction of critical rock mass state by the mine seismic data *J. Min. Sci* Vol 42 No 6 pp 523–529

[8] Lobanova TV, Lindin GL, Lobanov SA and Trofimova OL 2016 Effect of strong production blst on rock deformation and ground surface infrastructure *Naukoem. Tekhn. Razrab. Ispolz. Mineral. Res.* No 2

[9] Kasakhara K 1981 *Earthquakes Mechanics* Cambridge University Press

[10] Shrepp BV 1996 Geomechanical control of thick rockburst-hazardous iron orebody mining by changing the geometry and shape of mined-out void *Dr Eng Dissertation* Novosibirsk (in Russian)

[11] Revuzhenko AF 2006 *Mechanics of Granular Media* Springer-Verlag Berlin Heidelberg

Estimate of rock mass stability in surface–borehole mining of high-grade iron ore

AS Sammal[1], SV Antsiferov[1], PV Deev[1] and SV Sergeev[2]

[1]Tula State University, Tula, Russia
[2]Belgorod National University, Belgorod, Russia

E-mail: [1]antsser@mail.ru, [2]serg@bsu.edu.ru

Abstract. Under consideration is the estimate of rock mass stability around underground openings generated as a result of hydraulic borehole mining of iron ore. The authors use analytical solutions of two plane elasticity problems on stress state of infinite media with the zone of weakening in the form of one or two circular holes, given initial stresses are set in the study domains.

Hydraulic mining by boreholes is a promising method to develop difficult-to-access hard minerals. The main process flows involved in hydraulic mining is drilling, washing-out of minerals by water jets and lift of pulp slurry on the surface [1]. The evident advantages of this technology are low capital costs, applicability in complicated ground conditions, high-rate automation of the mining process and weak environmental impact.

The wide use of the hydraulic mining technology is mainly impeded by such factors as low actual productivity of holes as against the design capacity and risk of ground subsidence, especially in mining at shallow depths. As a consequence, hydraulic mining method is currently at the stage of trial, with low production outputs [2].

The most amplitudinous example of application of the hydraulic borehole mining method is iron ore production in the Shamrai site of Bolshe-Troitsk deposit in the Belgorod Region. Over the period from 2006 to 2013, eight boreholes produced round two thousand tons of ore, which resulted in generation of cylindrical cavities filled with pressure water at a depth of 500–700 m.

Upon completion of mining, a question arose on expediency of backfilling of the mined-out voids in order to prevent the ground surface from considerable subsidence and damage of water-tight strata. With this end in view, it was decided to assess stability of surrounding rock mass around excavations.

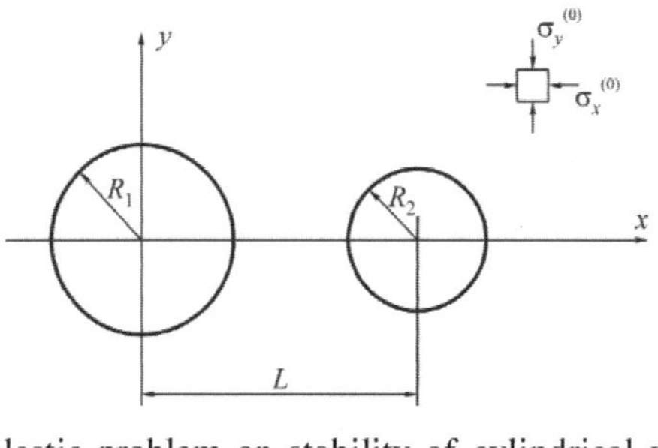

Figure 1. Schematic of an elastic problem on stability of cylindrical voids in iron ore–rock mass. Explanation is in the text.

The set problem had two stages: assessment of stability of walls in two cylindrical voids generated by the neighbor production boreholes and estimation of strength of a thick limestone bed overlying a layer of loose iron ore.

The first stage task used the known solution of plane elasticity on stress state of an infinite linearly deformable medium with two circular zones of weakening that modeled mined-out voids (Figure 1).

The stability assessment used the approach put forward in [3] and based on construction of conditional zones of inelastic strains, i.e. zones where the Mohr–Coulomb strength criterion holds true

$$\left(\sigma_\theta^2 - \sigma_r^2\right) + 4\tau_{r\theta}^2 > \left(\sigma_\theta + \sigma_r - 2C\,\mathrm{ctg}\varphi\right)^2 \sin^2\varphi,$$

where σ_r, σ_θ, $\tau_{r\theta}$ are the estimated stresses of rocks surrounding the voids (tension stresses are assumed positive); C is the cohesion, MPa; φ is the internal friction angle.

The operations on assessment of rock mass stress state and delineation of conditional zones of inelastic strains involved the input data: radii of the cylindrical voids (determined based on the ore production output under assumption that the voids were cylinders with the constant radii) $R_1 = 6.0$ m and $R_2 = 7.0$ m; spacing of the voids $L = 45$ m; strength characteristics of rocks (in the weakest bed, considering the structural weakening coefficient) $C = 1.6$ MPa, $\varphi = 30°$; initial stresses in the horizontal plane $\sigma_x^{(0)} = \sigma_y^{(0)} = -14.3$ MPa.

The calculations have shown that the mutual effect of the voids is insignificant, and the maximum size of the inelastic strain zone is not more than 2.2 m. Based on the classification of stability of excavations depending on the size of the conditional inelastic strain zones [3], a conclusion has been drawn that iron ore mass is insufficiently stable near the test voids and collapse is quite probable.

The second stage calculation used another approach to rock mass stability assessment. Under analysis was the stress state of a thick limestone bed overlying the iron ore bed where hydraulic mining was carried out. It was assumed that even in case of instability of the boreholes, the limestone bed would be sufficiently strong to eliminate probable subsidence.

The estimate of the limestone stability assumed that the iron ore bed contained a spherical void that touched limestone at the top of the bed. To find the influence of the void on the stress state of the limestone layer, a plane elasticity problem illustrated in Figure 2 was considered.

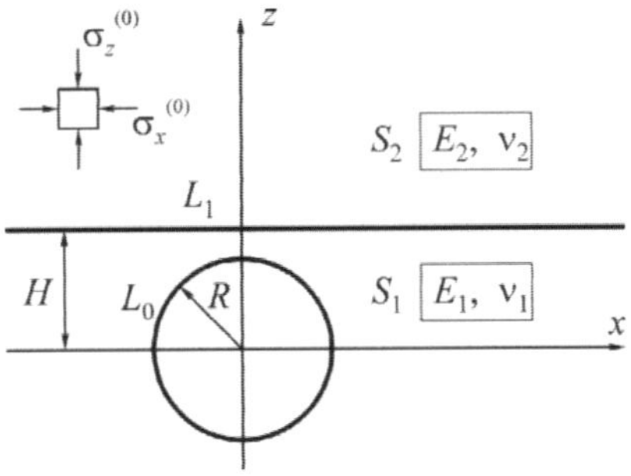

Figure 2. Schematic of an elastic problem on stress state of contiguous media with a circular zone of weakening in one of the media. Explanation is in the text.

In this case, a semi-infinite medium S_1 with an elasticity modulus E_1 and Poisson's ratio ν_1, weakened by a hole with a radius R, models the iron ore bed, and the medium S_2 with E_2 and ν_2 models the bed of limestone. In the media S_1 and S_2, the field of initial stresses is uniform and nonequicomponent. At the horizontal interface of S_1 and S_2, the conditions of continuity of the stress and displacement vectors are fulfilled, and the hole boundary is free of external forces.

It follows from the above formulation of the problem that the spherical cavity is replaced by the horizontal cylindrical hole, and the calculated stresses in limestone appear somewhat overestimated as a consequence. This assumption is applicable in actual calculations for the resultant error is included in the safety margin of mine excavations.

The problem was solved using the theory of functions of complex variable, Kolosov–Muskhelishvili integrated potentials [4], analytical extensions of these potentials, being regular in the lower semi-plane outside the hole, across the contact line, to the upper semi-plane [5], properties of integrals of the Cauchy type and complex series.

The assessment of stresses in limestone and the delineation of conditional zones of inelastic strains included the input data: E_1 = 1400 MPa, ν_1 = 0.33, E_2 = 59000 MPa, ν_1 = 0.30, C = 4.4 MPa, φ = 40°. The void radius determined based on the output of iron ore produced by the hole of the maximum capacity was R = 23 m.

The initial stress field due to the dead weight of rocks was assumed uniform and equicomponent (hydrostatic). The estimated initial stresses $\sigma_x^{(0)} = \sigma_z^{(0)}$ = 13.1 MPa. The distance from the hole center to the interface was assumed approximately equal to the hole radius, $H \approx R$.

The stress distribution in limestone at the interface is depicted in Figure 3. Figure 4 illustrates variation in normal stresses across the thickness of the limestone bed. The vertical coordinate h is counted from the iron ore and limestone interface.

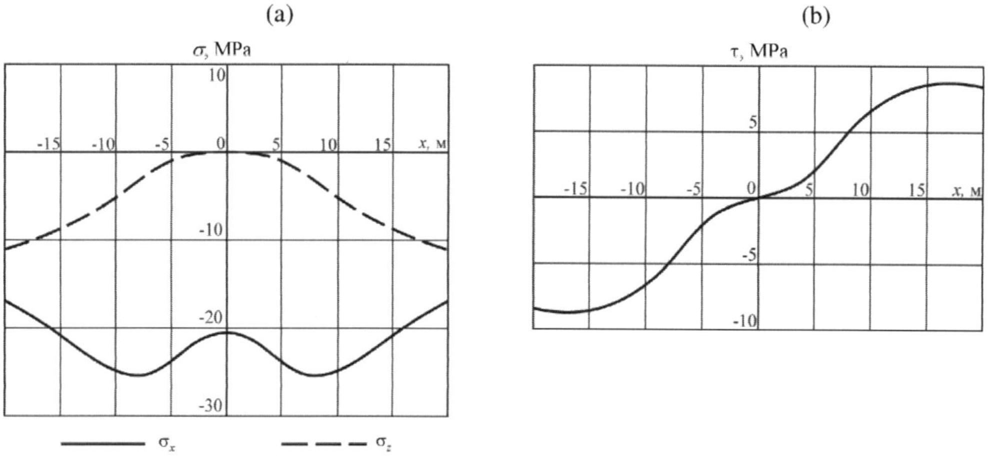

Figure 3. Distribution of (a) normal and (b) shear stresses at the interface of the two beds.

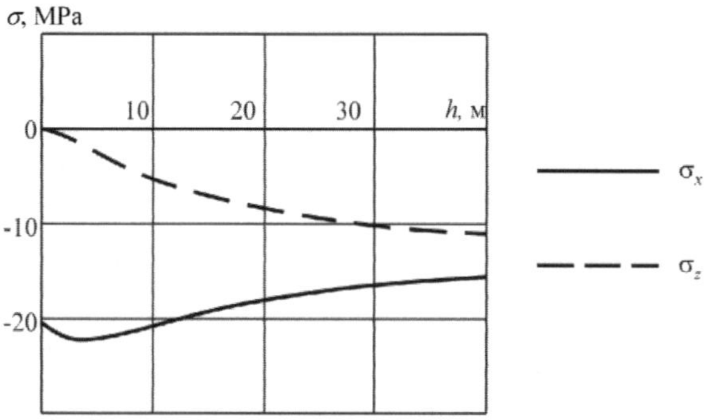

Figure 4. Variation in the normal stresses height-wise the limestone bed.

After the analyses of the resultant stress distribution in the limestone bed, the conditional zone of inelastic strains is localized at the limestone and hole contact and has a shape close to a circle, with a width of 12.2 m and a height of 1.6 m.

Since the maximal vertical size of potential damage zone is much less than the limestone bed thickness (approx 60 m), it is possible to conclude that the limestone bed in the vicinity of the production holes possesses sufficient stability and prevents from the ground surface subsidence and from disintegration of water-tight stata.

It is noteworthy that analytical calculations are currently applicable in solving of geomechanical problems of the types discussed in this paper only with considerable simplifications which reduce reliability of the obtained results.

References

[1] Arens VSh et al 2007 *Hydraulic Borehole Mineral Mining: Educational Aid* Moscow: Gronaya Kniga (in Russian)

[2] Boyarko GYu 2004 Hydraulic borehole mining *Metally Evrazii* No 1 pp 62–64

[3] Bulychev NS 1994 *Mechanics of Underground Structures: University Textbook* Moscow: Nedra (in Russian)

[4] Muskhelishvili NI 1966 *Somebasic Problems of Mathematical Elasticity Theory* Moscow: Nauka (in Russian)

[5] Aramanovich IG 1955 Stress distribution in an elastic semi-lane weakened by a cicular hole *Doklady AN SSSR* Vol 104 No 3

Evolution of microseismicity parameters depending on geomaterial deformation stages

VI Vostrikov[1], OM Usoltseva[1], PA Tsoi[1,2] and VN Semenov[1]

[1]Chinakal Institute of Mining, Siberian Branch, Russian Academy of Sciences, Novosibirsk, Russia
[2]Novosibirsk State Technical University, Novosibirsk, Russia

E-mail: [1]vvi@mail.ru

Abstract. The paper describes the laboratory tests on deformation of specimens made of artificial geomaterials simulating rock mass composed of two alternating beds of different hardness and different bedding angles. In the course of loading, stresses, strains and microseismic emission signals are continuously and concurrently recorded. The analysis of the time–deformation curve is reflective of the step-wise behavior of deformation and allows determining features of change in parameters of microseismic signals at various deformation stages.

Propagation of deformation and failure in loaded rock mass initiates various nature fields: deformation, microseismicity and electromagnetic emission, temperature. Some recent researches focus on the analysis of deformation and strength characteristics of rocks in integration with signals of microseismic emission that is assumed an analog to seismicity in lab tests [1–5].

It is of the current concern to find deterministic connection between the parameters of different physical nature fields as it is helpful in prediction of rock mass damage (rock falls, rock bursts, earthquakes).

This study was aimed to analyze interaction and cross-effect of parameters of different nature fields: deformation, stresses and signals of microseismic emission (MSE) in specimens under loading of two types: uniaxial loading of geomaterial to failure; fluid-fracturing of equivalent geomaterial for prediction of initiation and growth of main fracture.

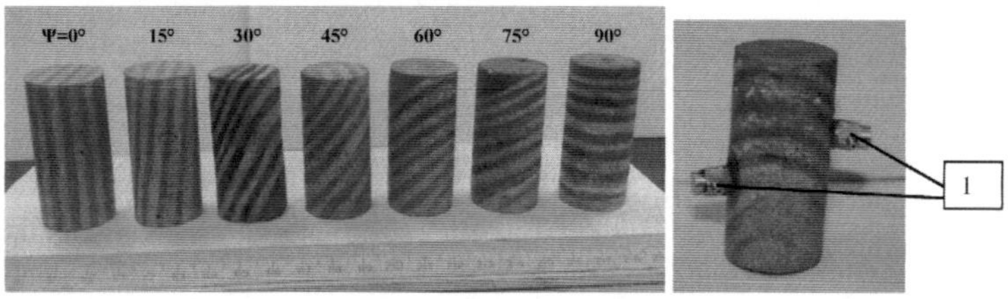

Figure 1. Specimens before the tests, bedding angle $\Psi = 0$, 15, 30, 0, 45, 60, 75, 90° (on the left); specimen with sensors to record signals of MSE (on the right).

I. In the first case, the test specimens were made of an artificial material to simulate layered rock mass composed of two alternate layers having different strengths and made of a mix of cement, sand, glue Neolit and water at varied ratios. The specimens were made as cylinders with the length of

60 mm and the diameter of 30 mm. The bedding angle (angle between the axis of the cylinder and the normal drawn to the plane of beds (isotropy) was $\Psi = 0$, 15, 30, 45, 60, 75, 90°. Figure 1 shows the pictures of specimens before the tests and a picture of a specimen with MSE sensors mounted.

The uniaxial compression of the specimens was applied up to failure at the movable jaw rate of 0.1 mm/min. During the tests, axial load and longitudinal and transversal deformation (change in the length and diameter, respectively) were continuously measured and filed using Instron sensors. Each tests series took not less than 3 specimens. The curves of the limit strength σ^{lim}, elasticity modulus E_{st} and bedding angle Ψ were plotted for each type of geomaterial (Figure 2).

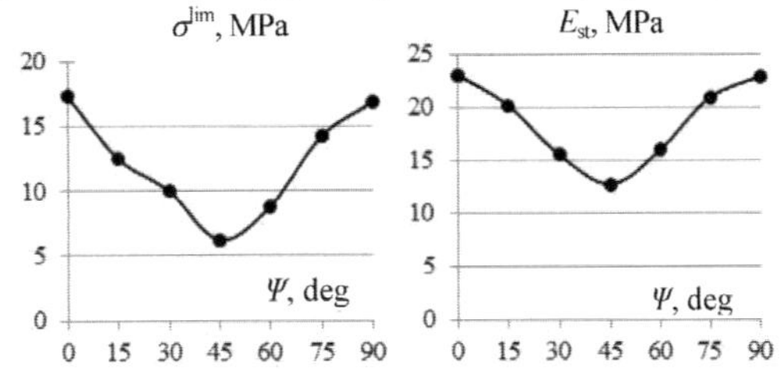

Figure 2. (a) Limit strength and (b) elasticity modulus of geomaterial under uniaxial compression versus bedding angle Ψ.

It follows from the test data analysis that the degree of anisotropy essentially influences deformation and strength characteristics of the specimens made of an artificial geomaterial with a layered structure. When $\Psi = 0$ and 90°, the limit strengths differ not more than by 3% and the elasticity modulus not more than by 4%, and the both values are higher at $\Psi = 0$ than at $\Psi = 90°$. The limit strength and elasticity modulus assume the lowest values when $\Psi = 45°$. In the range between 0 and 45 deg, the values of the discussed characteristics gradually decrease at the ratios of maximum/minimum strength limit and elasticity modulus of 2.8 and 1.8, respectively.

Concurrently with the measurement of stresses and deformation, MSE signals were recorded in the test. Below, the authors analyze MSE signals in the specimens with the highest difference in the strength and deformation characteristics, i.e. the specimens with the bedding angles of 45 and 90 deg.

Figures 3a and 3b present $\sigma/\sigma^{\text{lim}}–t$ curves for specimens nos. 4 and 9 with the bedding angles of 90 and 45 deg, respectively, where σ is the current stress, σ^{lim} is the uniaxial compression strength limit, t is the time, s. Specimens nos. 4 and 9 have $\sigma^{\text{lim}} = 17.3$ and 6.2 MPa, respectively. The stress and time curve is reflective of the step-wise behavior of deformation

Figures 3c and 3d show $\sigma/\sigma^{\text{lim}}–t$ curves for specimens nos. 4 and 9 from Figures 3a and 3b in more detail, at deformation stage 3. The dashed lines and figures point at the moments of MSE recording. Figures 3e and 3f give $\sigma/\sigma^{\text{lim}}–t$ curves for the specimen with the bedding angle of 45 deg for the time intervals 0–350 s and 350–500 s, respectively, for the better observability.

Figures 4a and 4b depict MSE signal recorded at point 1 in Figure 3e, and Figure 5—at point 2 in Figure 3f.

At the first deformation stage between zero load and $\sigma/\sigma^{\text{lim}} = 0.2–0.25$, the specimens undergo compaction and the elasticity moduli have lower values than in the subsequent stage, probably, due to damage of small heterogeneities at the faces of the specimens, or owing to closure of pores, micro-joints etc. At deformation stage 2 (elastic deformation) between $\sigma/\sigma^{\text{lim}} = 0.2–0.25$ and $\sigma/\sigma^{\text{lim}} = 0.55–0.65$, the elasticity moduli grow and become steady-state. MSE generated at this time has the acceleration of 7.0 m/s^2 (Figure 4a) and a wide frequency band (Figure 4b).

Deformation stage 3, from $\sigma/\sigma^{\text{lim}} = 0.55–0.65$ to $\sigma/\sigma^{\text{lim}} = 1$, features an essential reduction in values of the elasticity moduli, the material undergoes restructuring and micro-defects appear. The stress–time curve has horizontal segments and segments where the stresses somewhat decrease though the specimens are subjected to "stiff" uniaxial compression at the jaw rate of 0.1 mm/min. according to [6], when $\sigma/\sigma^{\text{lim}} = 0.5$, deformation is essentially nonuniform, with localization zones that shape a main fracture under the continued increase in the loading up to the limit strength value. At $\sigma = \sigma^{\mathrm{lim}}$ the stress–time curve is flattened, the values of σ decrease later on, and the main fracture is generated in the specimen.

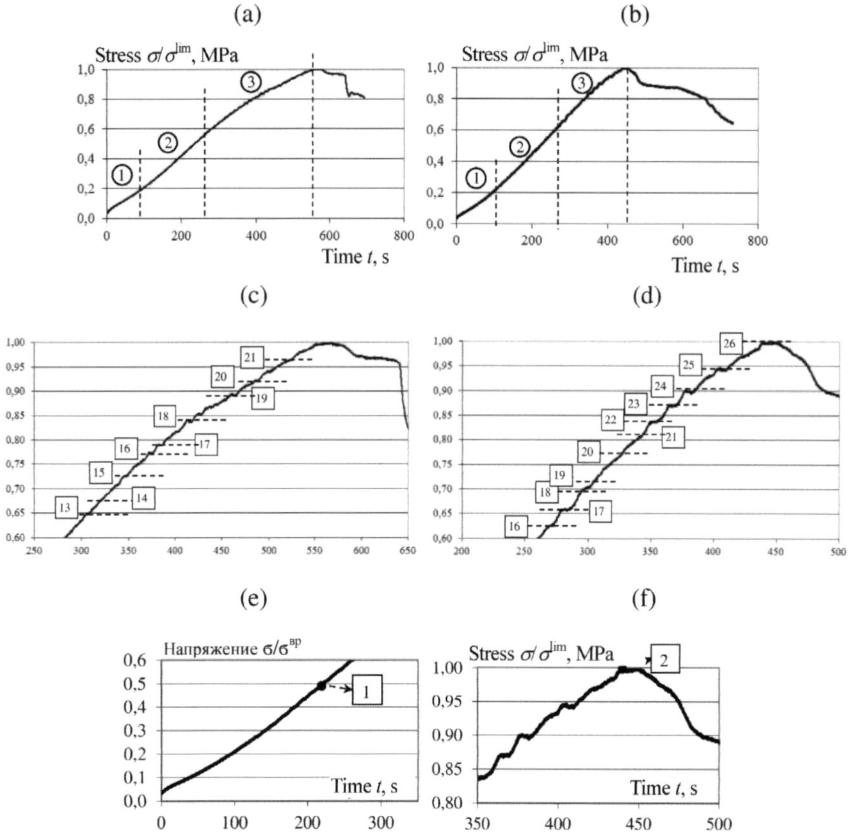

Figure 3. Stress–time curves for specimens (a) no. 4 and (b) no. 9; the same for specimens (c) no. 4 and (d) no. 9 at deformation stage 3 with MSE recording (figures and dashed lines); the stress–time curve for the specimen with the bedding angle of 45 deg in the time intervals of (e) 0–350 s and (f) 350–500 s. Squared 1 and 2 mark the times of MSE recording.

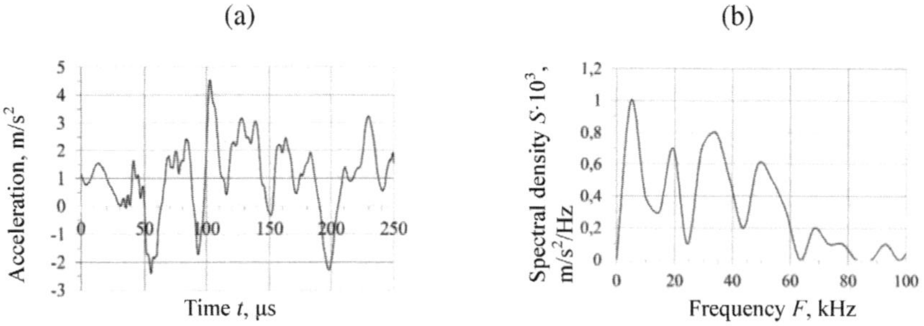

Figure 4. (a) and (b) MSE signal recorded at point 1 in Figure 3a.

Later on the MSE signal amplitude grows (Figure 5a) and its acceleration reaches 46.0 m/s^3. Frequencies shift to lower values of the frequency spectrum, and the spectrum becomes narrower (Figure 5b).

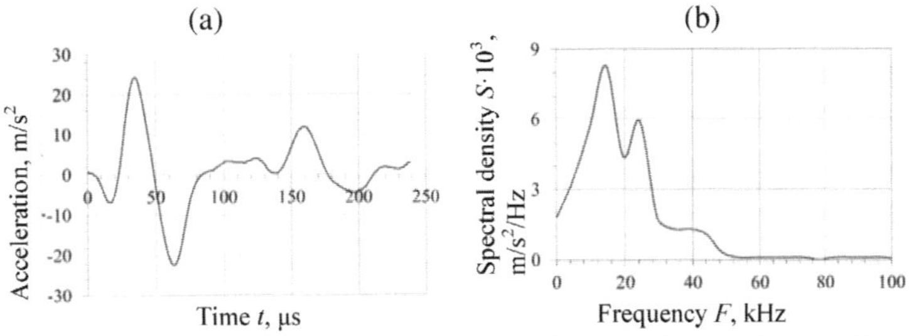

Figure 5. (a) and (b) MSE signal recorded at point 2 in Figure 3b.

The revealed mechanisms are most pronounced in the specimens with the bedding angle of 45 deg, probably, due to stronger deformation of these specimens, and re-structure of the geomaterial and initiation of new defects is more intensive in this case.

II. Oparin et al [6] studied the processes of deformation and failure in a material during its fluid-driven fracturing using three methods: speckle measurement of microstrains, acoustic emission and high-speed filming. The physical model was a parallelepiped 200×167×145 mm with an internal hole for fluid fracturing. During the tests, viscous fluid (plasticine) was fed in the hole at a constant flow rate.

The researcher analyzed the test results from various standpoints and drawn some conclusions:

—from the standpoint of time evolution of microstrains on the side surface of the specimen, in the area where the fluid-drive fracture reaches the surface. It was found that the field of microstrains of the material changed as the fracture grew. This means feasibility of studying internal destructive processes in a material by recording microstrains on its surface;

—from the viewpoint of time change in acoustic emission. It was found that MSE signal of the largest amplitude and frequency was recorded at the moment of the fracture initiation. The fracture propagated step-wise. When the fracture reached the material surface, strong low-frequency signal was recorded;

—in the context of interaction of these physical fields with the processes of the fluid fracture growth using the test video film.

Conclusion

The features revealed in the behavior of deformation and strength characteristics and MSE signals are reflective of clear inter-relation between them, which allows judging on destructive processes inside a test object. The research findings are useful for early prevention of tectonic activity and in monitoring of jointing in rock mass.

References

[1] Rizz E and Hahne P 2004 *Int. J. Plasticity* Vol 20 No 1

[2] Fukushima Y, Nishizawa O, Sato H and Oktake M 2003 Laboratory study on scattering characteristics of shear waves in rock samples *Bulletin of the Seismological Society of America* Vol 93 No 1

[3] Labuz JF, Cattaneo S and Chen LH 2001 Acoustic emission at failure in quais-brittle materials *Construction and Building Materials* Vol 15

[4] Damaskinskaya EE, Kadomtsev AG and Rasskazov IYu 2013 Detecting area of an intiating failure source based on deviation from the Guttenberg–Richter law *Proc. IX Int. Workshop Physical Bases of Rock Failure Prediction* Irkutsk (in Russian)

[5] Lementueva Russian Academy of Sciences , Bubnova NYa and Treusov AV 2013 Dynamics of main fracture growth in rock samples *Proc. IX Int. Workshop Physical Bases of Rock Failure Prediction* Irkutsk (in Russian)

[6] Oparin VN, Vostrikov VI, Usoltseva OM, Tsoi PA and Semenov VN 2015 Measuring equipment and test bench to control evolution of acoustic–deformation and heat fields insolids under failure by fluids *J. Min. Sci.* Vol 51 No 3 pp 624–633

Effects of structure on deformation and strength characteristics of transversely isotropic man-made geomaterials

OM Usoltseva[1], PA Tsoi[1,2] and VN Semenov[1]

[1]Chinakal Institute of Mining, Siberian Branch, Russian Academy of Sciences, Novosibirsk, Russia
[2]Novosibirsk State Technical University, Novosibirsk, Russia

E-mail: [1]usoltseva57@mail.ru

Abstract. The laboratory tests on uniaxial and triaxial (Karman scheme) compression of bedded specimens (made of equivalent man-made geomaterial, meta-siltstone and shale) has allowed deriving relations between the strength and deformation characteristics and the bedding angle of the specimens. The elasticity and strength are assessed in accordance with the theoretical model by Salamon–Tien and the Hoek–Brown failure criterion. For the bedded geomedia (man-made geomaterial), the Salamon–Tien model yields a satisfactory estimate of the elastic characteristics (elasticity modulus, Poisson's ratio). Based on the use of the Hoek–Brown criterion, the authors have derived a strength parameter independent of the lateral pressure.

Alternating strata of overburden rocks possess considerably anisotropic physical properties depending oт structure and texture of material, bedding angle, thickness and strength of strata, their geometry etc. The review of the recent publications shows that the researchers pay attention to bedded rock mass and to development of deformation processes in them [1–5]. It is highlighted in the studies that the angle of bedding has much influence on deformation and strength characteristics of rock mass. The research keeps relevancy as there are very many kinds of rocks.

This study describes experimental testing of variation in deformation and strength characteristics of a stratified geomaterial and rocks (meta-siltstone and schist) under uniaxial compression and triaxial compression by epy Karman scheme.

The tests used Instron 8802 hydraulic press with the loading rate preset as yoke displacement. Deformation and strength characteristics of a geomaterial were estimated in accordance with the ASTM standards. The geomaterial specimens were manufactured as alternation of two layers of different composition:

layer 1: sand—30 g, cement—10 g, Neolit glue—4 g, water—2.5 g;

layer 2: sand—30 g; cement—5 g, Neolit glue—3.5 g, water 2.5 g.

The angle of bedding (angle between the axis of a cylindrical specimen and the normal to the plane of beds—isotropy) was $\Psi = 0, 15, 30, 45, 60, 75, 90°$. Specimens made of different materials before the tests are shown in the pictures in Figure 1.

The uniaxial and Kaman's triaxial compression tests involved lateral pressures of 3 and 6 MPa (not less than 3 specimens per test series). For each material, the curves of the limit strength σ_1, deformation modulus E, transversal deformation ν and the bedding angle Ψ were plotted (Figures 2–4).

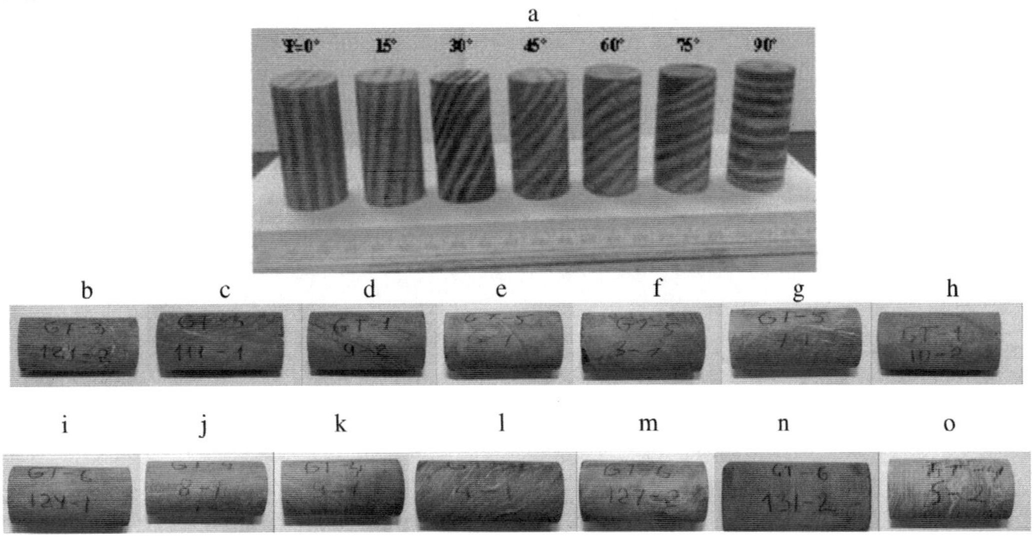

Figure 1. Pictures of specimens before the tests: (a) man-made geomaterial with $\Psi = 0°$, 15°, 30°, 0°, 45°, 60°, 75°, 75°, 90°; (b)–(h) siltstone with $\Psi = 0°$, 15°, 30°, 0°, 45°, 60°, 75°, 90°, respectively; (i)–(o) schist with $\Psi = 0°$, 15°, 30°, 0°, 45°, 60°, 75°, 90°.

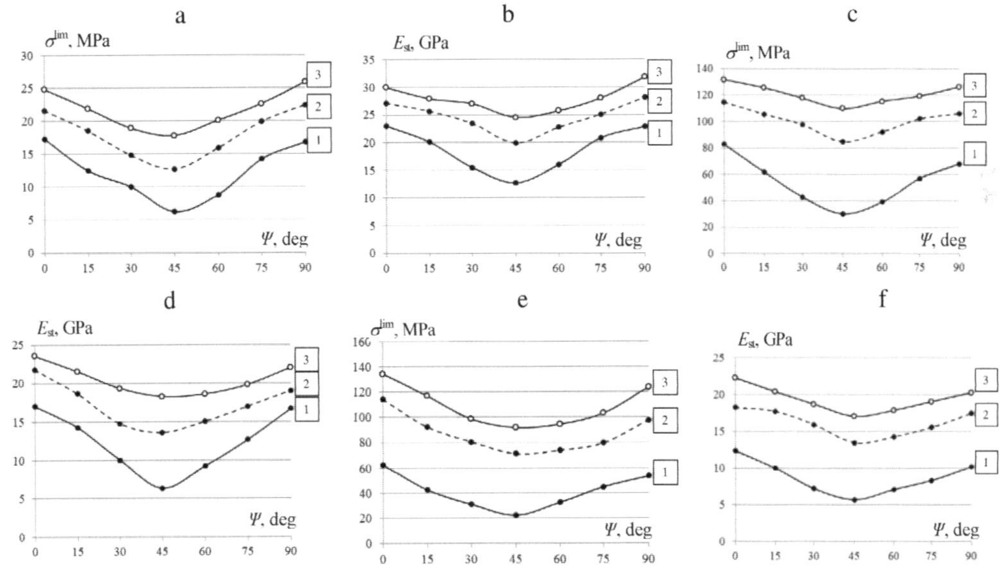

Figure 2. Limits strengths and deformation modulus versus bedding angles Ψ: (a) and (b) man-made geomaterial; (c) and (d) meta-siltstone; (e) and (f) schist under uniaxial compression (1), triaxial compression with the lateral pressure of 3 MPa (2) and 6 MPa (3).

The test data analyses show that the stratification essentially affects deformation and strength characteristics of the man-made material and rock specimens both under the uniaxial and triaxial compression. The strength is the lowest when $\Psi = 45°$ in all tests. The ratio of the strength limits at $\Psi = 90°$ and 0° makes:

Man-made geomaterial—2.79 under uniaxial compression,
1.78 under triaxial compression by the lateral pressure of 3 MPa and
1.45 under triaxial compression with the lateral pressure of 6 MPa;
Meta-siltstone— 2.74, 1.35 and 1.2;
Schist— 2.82, 1.6 and 1.46.

a b c d

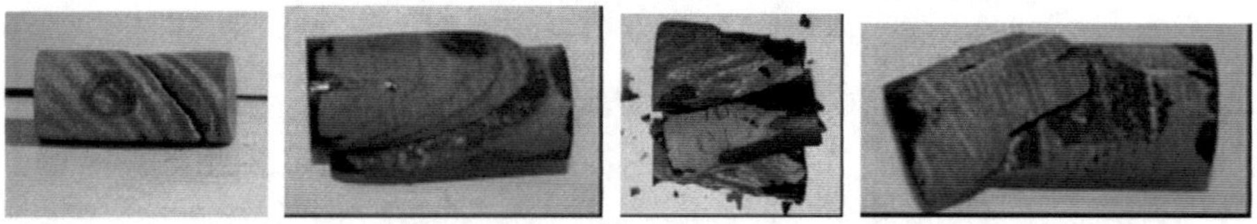

Figure 3. Specimens after the tests: $\Psi = 45°$—(a) man-made geomaterial; (b) meta-siltstone; $\Psi = 90°$—(c) meta-siltstone, (d) schist.

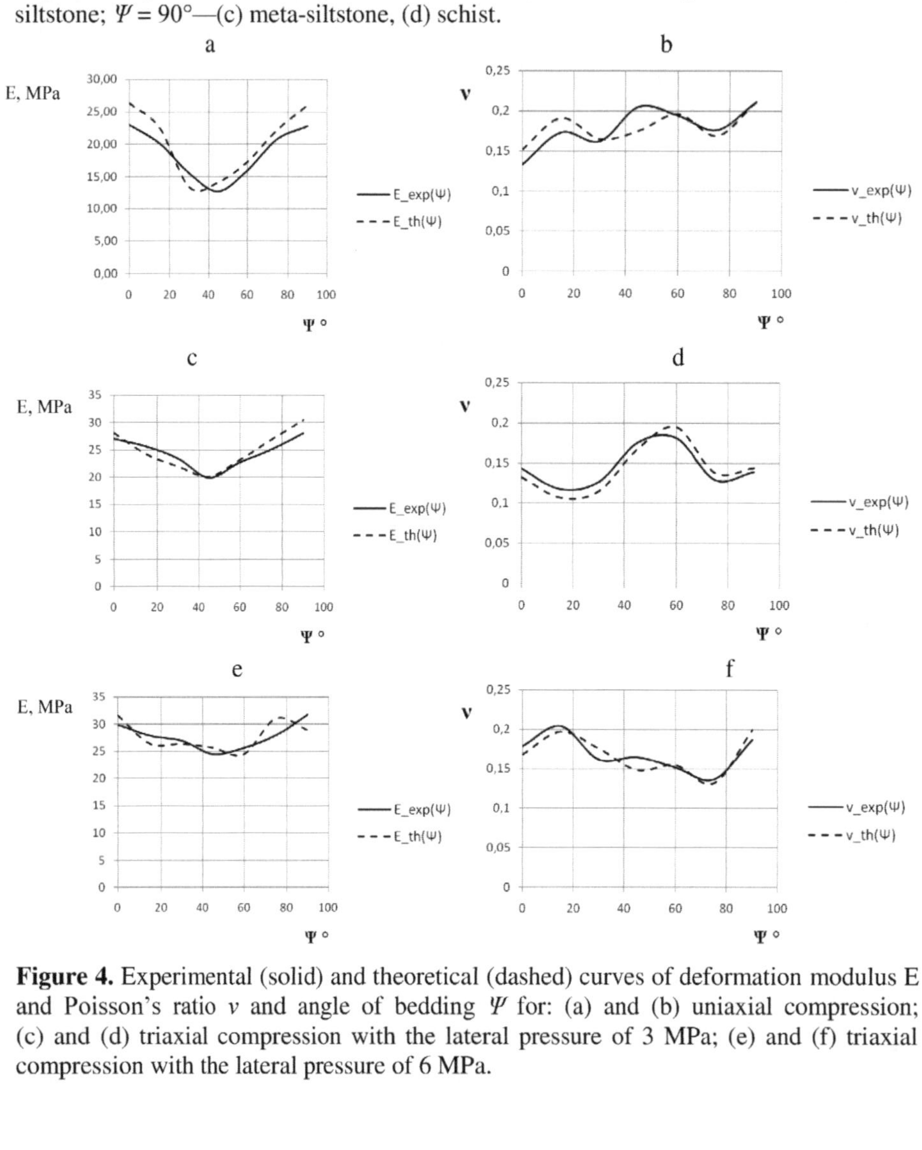

Figure 4. Experimental (solid) and theoretical (dashed) curves of deformation modulus E and Poisson's ratio ν and angle of bedding Ψ for: (a) and (b) uniaxial compression; (c) and (d) triaxial compression with the lateral pressure of 3 MPa; (e) and (f) triaxial compression with the lateral pressure of 6 MPa.

The E–Ψ curves behave the same way as with the limit strength. The deformation modulus acquires the least value when $\Psi = 45°$ in all tests. The ratios of the deformation moduli at $\Psi = 90°$ and 0° make:

Man-made geomaterial—1.81 under uniaxial compression,
1.41 under triaxial compression by the lateral pressure of 3 MPa and
1.29 under triaxial compression with the lateral pressure of 6 MPa;
Meta-siltstone— 2.68, 1.6 and 1.29;
Schist— 2.17, 1.35 and 1.31.

No clear regular pattern of change due to the bedding angle is observed for the coefficient of transversal deformation.

After examining damage of the test specimens, the conclusion is drawn that at $\Psi = 90°$ the damage zone contains a few main cracks; i.e. there is the longest line of fracture, which is consistent with the maximum values of the limit strength. When $\Psi = 45°$ failure propagates alone a line inclined at an angle of 35–55° to the specimen axis, as a rule (Figure 3).

The experimental characteristics of elasticity of the man-made geomaterial were compared with the theoretical values obtained using the Tien–Salamon equation. Assuming the deformation modulus and the transverse deformation coefficient as the elasticity modulus and Poisson's ratio, respectively, allows their assessment by the Tien–Salamon theory [6, 7]

$$\nu = \frac{\sum_{i=1}^{n} \frac{\lambda_i \nu_i E_i}{1-\nu_i^2}}{\sum_{i=1}^{n} \frac{\lambda_i E_i}{1-\nu_i^2}}, \quad (1)$$

$$E = \left(1-\nu^2\right)\sum_{i=1}^{n} \frac{\lambda_i E_i}{1-\nu_i^2}, \quad (2)$$

$$\lambda_i = \frac{H}{\sqrt[3]{V_i}}, \quad (3)$$

where λ_i is the dimensionless coefficient of volume for an i-th layer; V_i is the volume of an i-th layer; H is the layer thickness; ν_i and E_i are Poisson's ratio and elasticity modulus of an i-th layer; ν, E are Poisson's ratio and elasticity modulus of the whole specimen.

The maximum error of the estimate was not higher than 14.5%, which proved applicability of the proposed theoretical model for satisfactory assessment of elastic parameters of layered geomaterials.

For the man-made geomaterial, meta-siltstone and schist, the strength characteristics $m_{(0°)}$, $m_{(90°)}$ as per the Hoek–Brown criterion [8, 9] were calculated using the formulas below

$$m_{(0°)} = \frac{(\sigma_{1(0°)} - \sigma_3)^2 - \sigma_{c(0°)}^2}{\sigma_3 \cdot \sigma_{c(0°)}}, \quad (4)$$

$$m_{(90°)} = \frac{(\sigma_{1(90°)} - \sigma_3)^2 - \sigma_{c(90°)}^2}{\sigma_3 \cdot \sigma_{c(90°)}}, \quad (5)$$

$$m_{(0°)} = \frac{\frac{2\cdot(\sigma_{1(90°)} - \sigma_3)}{(\sigma_{1(30°)} - \sigma_3)} - \frac{1}{8} - \frac{(\sigma_{1(90°)} - \sigma_3)}{(\sigma_{1(0°)} - \sigma_3)} \cdot \frac{9}{8}}{\frac{9}{16} \cdot \frac{(\sigma_{1(90°)} - \sigma_3)}{(\sigma_{1(0°)} - \sigma_3)}}, \quad (6)$$

where $\sigma_{c(0°)}$, $\sigma_{c(90°)}$ are the limit strengths under uniaxial compression and bedding angles of 0° and 90°; $\sigma_{1(0°)}$, $\sigma_{1(30°)}$, $\sigma_{1(90°)}$ are the limit strengths under triaxial compression with the lateral stress σ_3 and bedding angles of 0, 30 and 90°. The table gives the calculation results.

Input and calculated criteria of failure by Hoek–Brown.

σ_3, MPa	$\sigma_{c(0°)}$, MPa	$\sigma_{c(90°)}$, MPa	$\sigma_{1(0°)}$, MPa	$\sigma_{1(30°)}$, MPa	$\sigma_{1(90°)}$, MPa	$m_{(0°)}$	$m_{(90°)}$	n	n_{av}
Man-made geomaterial									
3	17.3	16.8	22.6	14.8	22.4	1.63	1.86	2.76	2.59
6	17.3	16.8	25.8	18.9	25.2	0.89	0.85	2.42	
Meta-siltstone									
3	82.9	68.2	114.7	97.7	106.1	22.53	29.21	1.46	1.39
6	82.9	68.2	131.8	117.7	126.3	18.0	24.0	1.32	
Schist									
3	62.5	53.4	114.1	80.5	97.7	45.0	38.18	12	2.06
6	62.5	53.4	134.4	98.6	123.7	33.54	34.33	2.01	

Finally, the accomplished experimental research confirms that deformation and strength characteristics of an anisotropic stratified geomedium greatly depend on the medium structure. Anisotropy of rocks should be taken into account to avoid significant errors in calculation of stress state of rocks. The theoretical model of Tien–Salamon offers a satisfactory estimate of elastic parameters of bedded geomedia (elasticity modulus, Poisson's ratio). The strength parameter n_{av} calculated using the Hoek–Brown criterion of failure is independent of the value of lateral pressure.

References

[1] Vervoort A, Min K, Konietzkyc H, Cho JW, Debecker B, Dinh Q, Frühwirt T and Tavallali A 2014 Failure of transversely isotropic rock under Brazilian test conditions *International Journal of Rock Mechanics and Mining Sciences* Vol 701

[2] Li W, Petrovich C and Pyrak-Nolte LJ 2009 The effect of fabric-controlled layering on compressional and shear wave propagation in carbonate rock *International Journal of JCRM* Vol 4 No 2

[3] Hakala M, Kuula H and Hudson JA 2007 Estimating the transversely isotropic elastic intact rock properties for in situ stress measurement data reduction: A case study of the Olkiluoto mica gneiss, Finland *International Journal of Rock Mechanics and Mining Sciences* Vol 44

[4] Ghazvinian A, Geranmayeh Vaneghi R, Hadei MR and Azinfar MJ 2013 Shear behavior of inherently anisotropic rocks *International Journal of Rock Mechanics and Mining Sciences* Vol 61

[5] Vervoort A, Min K, Konietzkyc H, Cho JW, Debecker B, Dinh Q, Frühwirt T and Tavallali A 2014 Failure of transversely isotropic rock under Brazilian test conditions *International Journal of Rock Mechanics and Mining Sciences* Vol 701

[6] Tien YM and Tsao PF 2000 Preparation and mechanical properties of artificial transversely isotropic rock *International Journal of Rock Mechanics and Mining Sciences* Vol 37 pp 1001–1012

[7] Salamon MDG 1968 Elastic moduli of a stratified rock mass *International Journal of Rock Mechanics and Mining Sciences* Vol 5 pp 519–527

[8] Tien YM, Kuo MC and Juang CH An experimental investigation of the failure mechanism of simulated transversely isotropic rocks *International Journal of Rock Mechanics and Mining Sciences* Vol 43 pp 1163–1181

[9] Tien YM and Kuo MC 2001 A failure criterion for transversely isotropic rocks *International Journal of Rock Mechanics and Mining Sciences* Vol 38 pp 399–412

The surface waves-based seismic exploration of soil and ground water

AS Serdyukov[1,2], AV Yablokov[1], GS Chernyshov[1] and AV Azarov[2]

[1]Institute of Petroleum Geology and Geophysics, Siberian Branch, Russian Academy of Sciences, Novosibirsk, Russia
[2]Chinakal Institute of Mining, Siberian Branch, Russian Academy of Sciences, Novosibirsk, Russia

E-mail: [1]serdyukovas@ipgg.sbras.ru; [2]antonazv@mail.ru

Abstract. We consider the method to study physical properties of soils. The new seismic exploration approach, based on new modification of refracted wave plus-minus method and multichannel analysis of surface waves (MASW), is proposed. We present the field data processing example. The ground water level and zones of erosion are detected.

1. Introduction

For geotechnical investigations and assessment of pitwall stability, the study of physical and mechanical properties of rocks and soils in situ is required. First and foremost, it concerns dispersive soils. It is possible to determine physical and mechanical properties of cohesive dispersive soils with the help of seismic exploration. According to the studies [1, 2] P- and S-wave velocities can be used for the geotechnical investigations. The primary properties defined by the regulations of the investigations are: soil moisture W, yield point Wm, soil density ρ, matrix density $\rho_{ск}$ and deformation modulus E_o. All mentioned properties can be determined with adequate accuracy from the velocity profiles of P- and S-waves [3]. Correlation formulas for typical cohesive dispersive soils (sandy loam, loams, clays, etc.) derived by regression analysis are presented in methodological recommendations [4].

For seismic shallow investigations spread line system with one-component sensors is typically used to register vertical oscillations on the daylight surface. Pulse percussion sources cause the oscillations. It is possible to induce and register the transverse oscillations with the suggested scheme, but it is inconvenient due to the following reasons. Thus, the low-velocity zone is characterized by sudden changes of the elastic properties and density as well as the contrasting refracting boundaries. It is difficult to assess the influence that complex medium features have on dynamic properties of seismic waves; therefore, predominantly kinematic methods are used for the wave studies [5–7]. As a rule, seismograms show the easily distinguished first arrivals of P-waves, including the arrivals of direct waves close to the source, and refraction waves at a distance from the source. It is complicated to employ reflected waves because they are rare distinguishable from the surface waves with larger amplitude [8].

The standard method of one-component data processing in geotechnical investigations is to restore velocity profile of P-waves using refraction method, P-waves velocities are estimated by Knopov charts [4]. In the classical algorithms [9] the depth of the refracting boundaries is determined, and velocities between layers are constant. On the basis of the data on head waves velocities distributions along a refracting boundary are reinstated. The direct wave data help to restore average velocities in near-surface layer. Practically there can be lateral changes of P-wave velocities, for example, changes due to moisture. When constructing the velocity profile using the classical method t_0' ("plus–minus" method) heterogeneity areas of hodographs correlating with the top layer are interpreted as the changes in the reflecting boundary depth [10]. The heterogeneity may be covered by lateral perturbations in velocity distribution of a P-wave in the top layer. In order to adjust the heterogeneity a priori information has to be employed. This paper discusses a typical case in which the boundary coincides with the ground water level, and only gradual changes of refraction depth is implacable. Therefore, it is necessary to even the boundary by equivalent velocity perturbations in the top layer. In this work we have implemented the described idea by means of the modified refraction method t_0'.

To build a velocity profile of S-waves we use multichannel analysis of surface waves (MASW) [11] instead of Knopov pattern. The main point of this approach is reduced to the construction of a stratified model based on the dispersion curves of the Rayleigh wave phase velocity.

2. Modified refraction method

Refracted wave t_0' method, known abroad as "plus–minus" method, is presented in the literature in details [10]. While using the method to process field data of shallow seismic investigations the form of the restored boundaries is usually not smooth with the small local depth drops. From physical standpoint these frequent changes are unlikely to happen, for example, the level of the upper groundwater may gradually change, but frequent small irregularities are not typical.

Let us consider the case when the velocity of the first layer, which is determined by the direct wave travel time, is not precise. In field measurements this situation may be the result of a number of reasons: the low frequency of the probing signal, low power of the near surface layer, seldom arrangement of the detectors, the irregularities in the bottom of the layer. Then, the perturbations of the refracting boundary δh depth and P-wave velocity variations in the upper layer δV_1 are connected as

$$\delta h = \frac{T_D^+ \delta V_1}{2\cos^3\theta}, \qquad (1)$$

where $T_D^+ = T_{AD} + T_{HD} - T_{AH}$ —expression composed of travel times of the head wave, T_{AD}, T_{HD}, T_{AD} — travel times between the corresponding points (Figure 1). Using formula (1) one can interpret minor departures from the depth smoothness of a bearing boundary, defined by the classical method t_0' as the velocity variations in the upper layer.

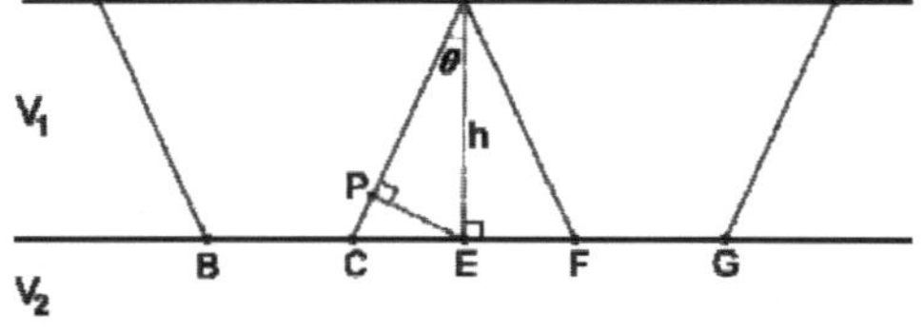

Figure 1. Linear scheme of seismic wave propagation.

The suggested method involves the following steps: firstly the classical method t_0' is employed, then, the found boundary is smoothed, and the difference between the initial and smoothed depths is inserted into the formula (1) to make the velocity in the upper layer more precise. Supposing that the perturbations are minor, the formula (1) is obtained due to linearization, and it is more efficient to use the formula in several steps (gradually smoothing the boundary). In accordance with this principle, the depth-gradient model is built for every move-out.

2. Multichannel analysis of surface waves

The next step of data processing is transition into spectral f-k domain. Two-fold Fourier transformation is applied to the seismogram in respect with a measuring points offset coordinate and with time. Then, there is transformation of f-k (1/s-1/m) figure into the dependence domain of Rayleigh wave phase velocity on temporal frequency V-f (m/sec—1/sec). There is a high level of signal/noise ratio due to the fact that not less than 2/3 of the energy is spent on the surface wave formation while oscillations are induced. The dispersive curves are obtained by setting the amplitude maximums in the observing frequency range depending on the seismic profile length, distance between the seismometer, seismogram recording length and sample spacing.

Re-establishment of the S-wave velocity profile is based on the adjustment of the layered one-dimensional model (of S-wave velocity) using the dispersive curve and repeated solving of the primal problem. Next, we implement interpolation of one-dimensional profiles in order to get 2D (or 3D) models of the geological cross section under the survey. The prime problem is solved in the one-dimensional model of the medium consisting of horizontal plane–parallel homogenous isotropic layers. Their phase velocity dependence on the frequency (dispersive curve) can be found analytically, by finding eigenvalues for displacement–stress vector [12]. We use a modified matrix propagator method, the so-called "method of reflection and transmission coefficients." Then, we search for the minimum of residual functional between the observed and calculated dispersive curves. Minimization is performed with the help of Levenberg–Marquardt algorithm, which is a modification of the Newton method [13].

3. Interpretation methods of velocity profiles and analysis of dispersive cohesive soil properties

It is possible to interpret medium boundaries comparing V_p and V_S. S-wave velocities, unlike P-wave velocities, do not undergo jumps at the groundwater boundaries [4]. If there are lithological boundaries above the ground water level V_p and V_S changes are observed. In case when lithological boundaries are below the ground water level there are jumps of V_S while there are subtle changes in V_p [4]. Therefore, it is very important to use relevant geological information as well as geoelectric data. Characterization of physical and mechanical properties of cohesive dispersive soils is based on correlation dependence given in methodological recommendations [4]. The presented formulas are applicable for the soils above the ground water level with less than 30% moisture. If the soil is greatly saturated, P-wave propagates along the liquid, and its velocity weakly depends on rock matrix properties. To calculate moisture W and matrix density of the rock $\rho_{ск}$ we have employed iterative updating procedure of values. At first, the rock density is estimated approximately $\rho_{ск} = 1{,}38 + 0{,}00033\, V_p$, where velocity is measured in m/sec, and density is measured in g/cm^3. In what follows empirical constants have the corresponding degrees. The calculated matrix density is used to measure the moisture

$$W = 79{,}06 - 0{,}00382\, V_p - 30{,}1 \frac{V_s}{V_p} - 28{,}91\, \rho_{ск}. \quad (2)$$

The moisture is measured in percent of the mass. Calculated by the formula (1) moisture is used to get a more precise value of the matrix density

$$\rho_{ск} = 2{,}34 - 0{,}736 \frac{V_s}{V_p} - 0{,}0245\, W. \qquad (3)$$

The specified matrix density is inserted into the formula (1) and etc. The conducted numerical experiments prove that for typical P- and S-wave values iteration process (1)–(2) converges in 10 iterations. The process is stopped if the absolute values of difference between the values in two last iterations do not exceed tenth of percent of the last iteration values.

An important physical parameter is the yield point W_m. It is moisture due to which soil state turns from plastic condition to liquid one. The yield point is calculated using the formula $W_m = 36,93 + 0,8W - 13,143\rho + 0,0429V_s$.

The calculated matrix density and yield point values allows estimating the deformation modulus

$$E_0 = 0.0549V_p + 0.666V_s + 145.4\rho_{ск} + 20,66(W_m - 27.8)(\rho_{ск} - 1.56) - 266.1 \text{, [kg/cm}^2\text{]},$$

characterizing compression (elastic and plastic parts) of soil in case of static (quasistatic) vertical loadings. It is possible to determine other characteristics [1, 4].

4. Field data processing results

Shallow seismic investigations have been carried out in the suburb of Novosibirsk on the 8th of October 2015 along the road Akademgorodok–Klyuchi, Kamenushka river. The distance between the detectors of the linear observing system is 1 m. One-component vertical seismic receivers with 10 Hz frequency are used. Source points are placed along the observation line at a 5 m distance from each other. To induce oscillations in all the points a 5 kg sledgehammer is used to hit a metal base. Figure 2 shows the constructed P wave velocity profile. At the 4 m-depth we have found the underground water line. The top of Figure 2 presents the velocity profile constructed using the well-known method t_0'. At the bottom of this figure, there is a velocity profile derived by the suggested modified method.

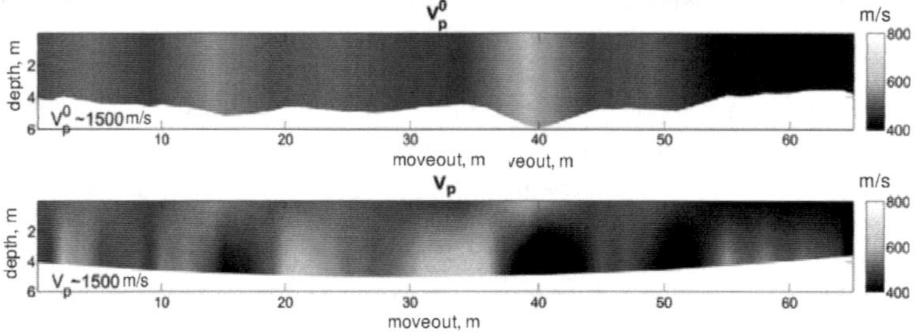

Figure 2. P-wave velocity profiles: at the top—obtained by the standard method; bottom—by the modified method.

Figure 3 shows the S-wave velocity profile resulted from the multichannel analysis of surface waves. Comparing S- and P-wave velocity profiles we have come to the conclusion that the observed line in Figure 3 marks the ground water level. Smoothness of the line is caused by the physical properties, and it shows the efficiency of the developed method.

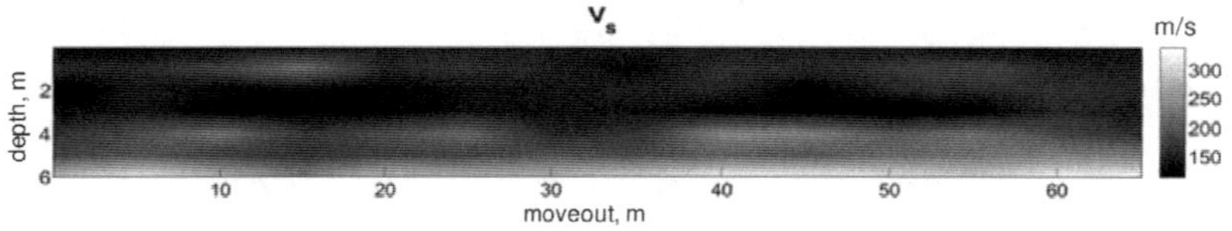

Figure 3. S-wave velocity profile.

Figure 4 introduces the restored soils properties distribution down to a depth of 3.5 m. At the top of the figure there is the ratio between the yield point and moisture. The values close to 1 are the indicators of transformation into liquid state in case of a slight moisture increase. At the bottom of the figure there is general deformation modulus characterizing soil compressibility. Areas with the low values of the modulus are more likely to be deformed. Areas of concern are settlements on the 12–24 m of the profile and quick grounds at 43–56 m of the profile. They correlate with the changes of properties in Figure 3. The obtained results are in agreement with the road observation which proves the effectiveness and efficiency of the suggested method.

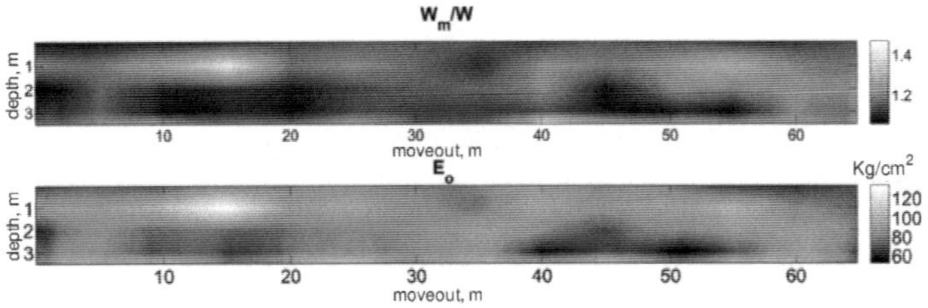

Figure 4. Soil properties distribution: top—ratio between the yield point and soil moisture; bottom—soil deformation modulus.

5. Conclusion

For geotechnical investigations we have developed a seismic method based on the concurrent processing of the first P- and S-wave arrival times. The suggested method based on correlating dependences makes it possible to restore physical and mechanical properties of cohesive dispersive soils and to determine areas subjected to deformation and erosion.

6. Acknowledgements

The research is supported by Russian President grant for young PhD scientists No. MK-7778.2016.5

References

[1] Pioro EV and Oshkin AN 2011 Interrelations of acoustic characteristics and physical and deformation properties of clay soils *MSU Vestnik. Geology* Vol 26 pp 71–74

[2] Goryanov NN 1992 Employment of seismic-acoustic methods in hydrogeology and engineering *Geology* Vol 13

[3] Trofimova VT 2005 *Soil Science* Moscow: Nauka (in Russian)

[4] Anikin OP 1976 *Methodological Recommendations on the Use of Seismic-Acoustic Methods for Studying Physical and Mechanical Properties of Soils*

[5] Kurlenya MV, Serdyukov AS, Duchkov AA and Serdyukov SV 2014 Wave tomography of methane pockets in coal bed *Journal of Mining Science* Vol 50 pp 3–9

[6] Serdyukov AS, Patutin AV and Shilova TV 2014 Numerical evaluation of the truncated singular value decomposition within the seismic traveltimes tomography framework *Journal of Siberian Federal University. Mathematics & Physics* Vol 7

[7] Serdyukov AS and Duchkov AA 2015 Hybrid kinematic-dynamic approach to seismic wave-equation modeling, imaging and tomography *Mathematical Problems in Engineering* 543540

[8] Kurlenya MV, Serdyukov AS, Azarov AV and Nikitin AA 2015 Numerical modeling of wavefields of microseismic events in underground mining *Journal of Mining Science* Vol 51 pp 689–695

[9] Goldin SV, Kiseleva LG anf Kurdyukova TV 1985 Interpretation of refracted wave godographs by king system in case of rugged daylight surface *Geolog. Geofiz.* Vol 12 pp 120–126

[10] Van Overmeeren RA 2001 Hagedoorn's plus-minus method: the beauty of simplicity *Geophysical Prospecting* Vol 49 pp 687–696

[11] Park CB, Miller RD and Xia J 1999 Multichannel analysis of surface waves *Geophysics* Vol 64 pp 800–808

[12] Lai CG and Rix GJ 1998 *Simultaneous Inversion of Rayleigh Phase Velocity and Attenuation for Near-Surface Site Characterization* Georgia Institute of Technology

[13] Solano CA 2013 *Two-Dimensional Near-Surface Seismic Imaging with Surface Waves: Alternative Methodology for Waveform Inversion* Mines ParisTech

Laboratory research of hydraulic fracturing with tangential loading of borehole wall

MV Kurlenya, AV Patutin*, LA Rybalkin, SV Serdyukov and TV Shilova

Chinakal Institute of Mining, Siberian Branch, Russian Academy of Sciences, Novosibirsk, Russia

E-mail: *andrey.patutin@gmail.com

Abstract. Under study is transverse fracturing of an organic glass block through secondary shearing stress applied to the borehole wall. To this effect, a system composed of a press sealer and a collet anchor manufactured in two options has been designed. It is shown than an anchor with a circular groove allows reducing breakdown pressure and enables effective transverse fracture at the borehole bottom.

Hydraulic fracturing method is a primary way to increase the effectiveness of borehole mining technologies employed in solid minerals mining in underground conditions. Taking into account previous experience one can conclude that it is practically useful to create in a borehole one or several parallel to each other transverse cracks. The hydraulic fracturing method can be employed to soften the hard coal bed top [1], to increase gas-recovery [2], to create impervious screens [3] and to extend effective area of thermal action and in-situ leaching on rock mass [4].

Well-known methods to create a hydraulic fracture across the borehole axis are based on employment of slot initiators formed in rock by jet or mechanical slot generators. The main drawback of the method is the necessity to cut in the rock mass a fracture initiator, which requires additional time and energy-consuming operation. To increase the effectiveness of the directional hydraulic fracturing in uncased borehole a slotless method is suggested. This method implies that energetically beneficial direction of the fracture propagation is set by anchors due to non-destructive changes of the stress condition of rock masses near the borehole [5]. Buoyant force of the working fluid pressure influencing the sealer is transmitted to the anchor's contact to the rock mass. It creates the additional tangential loading on the borehole walls, which causes the rocks tension along the borehole axis and formation of a transverse crack. This work is aimed at the experimental test of the slotless method of directional hydraulic fracturing in laboratory setting.

To implement the described approach we have developed a directional fracturing tool general view of which is shown in Figure 1.

The tool consists of a collet anchor *1*, pressure sealer *6*, tube *8*, screw nut *10*, draft *9* with the inner channel *5* and a hydraulic output *11*. The anchor is made in two variants: one with the ring groove *b*, forming the sharp indenter point *a* and another—without the groove. As for the first variant coupling of the anchor and the borehole walls takes place along the indenter impressing in the model's materials with the contact area 2–4 times less than the borehole's cross section. In the second variant the anchor's coupling area exceeds the cross section in 2–4 times.

When the experiments are being performed the tool is inserted all the way in the borehole bottom, after that, the draft *9* is screwed in the anchor *1* and placed in the operating position. Then, the sealer

6, plate *7*, tube *8* and screw nut *10* are set, and the fracture interval is sealed. The output *11* is connected with a hydraulic station by a high-pressure hose. The working fluid is injected in the fracture between the borehole bottom and the sealer through the input *11* and canal *5*

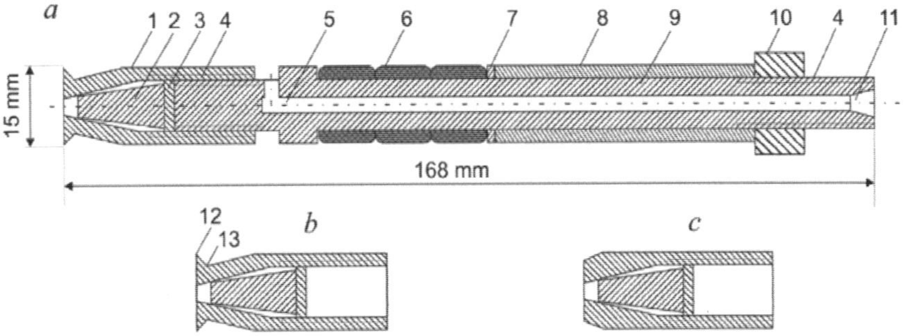

Figure 1. Plain view of the hydraulic fracturing tool: *a*—hydraulic fracturing tool: 1—wedge anchor; 2—cone; 3—pad; 4—screw joints; 5—fluid channel; 6—sealer; 7—washer; 8—tube; 9—draft; 10—screw nut; 11—hydraulic input; b—improved wedge anchor: 12—indenter; 13—ring groove; c—standard wedge anchor

The fluid pressure in the isolated interval (*P*) extrudes the tool from the borehole with the power $\pi R^2 \cdot P$, where R – borehole's radius. The power is transmitted to the anchor creating the tangential loading on the anchor contact with the borehole. The loading and the fluid pressure at the borehole bottom cause the model's material stretching along the borehole axis [6]. In the variant with the indenter the loading is localized along the output line formed by the edge crack indenter, and its value approaches $PR/2$. When the standard anchor is employed the tangential loading is distributed along the area $2\pi R \cdot L_a$, where L_a is the length of the anchor's contact with the borehole wall ($L_a \approx 1.5R$). In this case the tangential loading does not exceed $P/2$, therefore, employment of the suggested approach is preferable [7].

Figure 2 shows the laboratory bench scheme to carry out the experiments. It consists of a model, hydraulic and measuring systems. The model is a 250×400×600 mm plexiglas slab with the holes (the so-called boreholes) drilled in it. Their diameter is 16 mm and depth is equal to 80–120 mm.

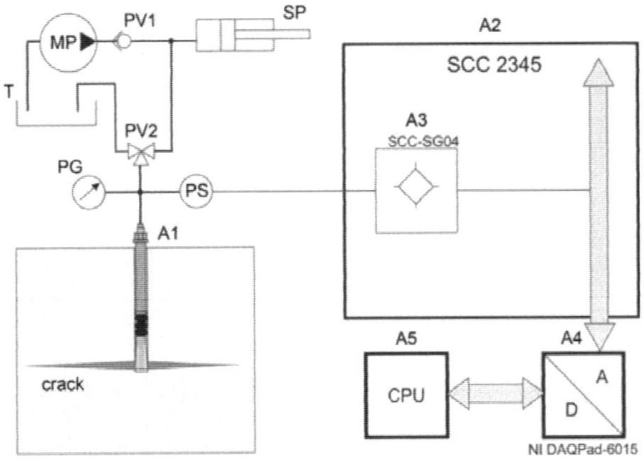

Figure 2. Laboratory bench scheme for the study of directional hydraulic fracturing.

The hydraulic system of the bench contains the hydraulic fracturing tool A1, shown in Figure 1, the high-pressure manual pump MP with the backflow prevention valve PV1 at the output, syringe pump SP for precise injection, valve PV2, pressure gauge PG, drain tank T and high-pressure connection hoses.

IOP Conf. Series: Earth and Environmental Science 53(2017) 012011 doi:10.1088/1755-1315/53/1/012(

The measuring system of the bench includes the pressure sensor PS, crate A2, equipped with the module A3, as well as analog-to-digital converter A4 and computer A5. There is a strain-gage pressure sensor LH-412/400 (Tenzopribor, Russia), able to measure the working fluid's rapidly changing pressure from 0 to 40 MPa in the frequency range 0-1500 Hz with an accuracy of ±0.8%. The sensor PS electric signal goes to the measuring module outputs of the equipment SCC (National Instruments, USA) set in the crate SCC-2345 (A2). In the work SCC-SG04) (A3) module is used to connect the bridge sensors. The output signal of the module SCC is transmitted to the input of the analog-to-digital converter NI DAQPad-6015 (A4) through the tool bus of the crate A2. Through the USB port the digitized data is transmitted to the computer A5, where the files are saved in txt format. Data collection and measuring system management is organized in LabView (National Instruments, USA).

The laboratory tests on the directional hydraulic fracturing are carried out according to the following methodology:

1) the fracture tool is installed in the prepared model, the hydraulic and measuring parts of the bench are attached;

2) the continuous recording of the sensor PS output signal with the discretization interval 1 m/sec;

3) fracturing interval is filled with the working fluid up to pressure 1.0 MPa;

4) the working fluid pressure is getting increased by syringe pump SP in the fracture interval until a crack is formed. The crack formation is recorded by the sudden drop of pressure on the pressure gauge PG. After that the working fluid is not supplied while the hydraulic system is kept closed. In case when the fracturing is performed with the help of the anchor without the ring groove a longitudinal crack is formed along the well (Figure 3a). If there is an indenter point a cross crack is formed (Figure 3b).

a b

Figure 3. Hydraulic fracturing cracks formed in experiments: a—transverse crack; b—longitudinal crack.

The laboratory test results on the hydraulic fracturing performed with the developed tool are presented in the table.

The averaged pressures of the plexiglas fracturing are: in case of the longitudinal hydraulic fracturing—21.71±5.09 MPa, in case of the cross hydraulic fracturing—16.01±4.39 MPa. Confidence intervals correspond with the possibility of 0.9. There is a significant difference in the fracture pressures in both fracture schemes. It is not connected with the characteristics of the medium as the experiments are performed on the one material. It is likely that the difference is caused by different quality of borehole boring as well as the influence of the industrial micro crack sizes on the plexiglas resistance to the hydraulic fracturing. The second reason is the possible difference in the boreholes' distances from the free surfaces. Thus, the longitudinal fracture in the area of the transverse crack appeared due to the abnormally low pressure 7.65 MPa. It is important to mention that when the fracturing is performed with the indenter a transverse crack is formed in all the experiments. The obtained data display the working efficiency of the directional hydraulic fracturing.

Results of laboratory tests on hydraulic fracturing cracks' orientation and pressure.

Fracture scheme	Crack orientation	Crack output at model's edge	Fracture pressure, MPa	Shut-in pressure, MPa	Crack diameter, mm
Anchor with indenter	longitudinal	no	13.44	2.99	91–104
			16.52	2.00	114–174
		yes	18.01	0	–
			21.13	0	–
			33.17	0	–
			20.37	0	–
			29.35	0	–
Anchor with indenter	transverse	yes	19.02	0	214
			14.22	0	–
			14.80	0	–

Acknowledgements
The research was carried out in Institute of Mining SB RAS and supported by the grant of the Russian Scientific Foundation, project № 15-17-00008

References

[1] Lekontsev YM Sazhin PV 2008 Application of the directional hydraulic fracturing at Berezovskaya Mine *J. Min. Sci.* 44 (3) 253–258
[2] Huang B Cheng Q Chen S 2016 Phenomenon of methane driven caused by hydraulic fracturing in methane-bearing coal seams *Int J Min Sci Technol* http://dx.doi.org/10.1016/j.ijmst.2016.05.042
[3] Kurlenya MV Shilova TV Serdyukov SV Patutin AV 2014 Sealing of coal bed methane drainage holes by barrier screening method *J. Min. Sci.* 50 (4) 814–818
[4] Vorobiev AE Chekushina TV Pogodin ML Odincova ES 2000 In situ leaching of gold in Russia *Min. Inf. Anal. Bull.* 12 45–48
[5] Shilova TV Serdyukov SV 2005 Protection of operating degassing holes from air inflow from underground excavations *J. Min. Sci.* 51 (5) 1049–1055
[6] Azarov AV Kurlenya MV Patutin AV Serdyukov SV 2005 Mathematical modeling of stress state of surrounding rocks around the well subjected to shearing and normal load in hydraulic fracturing zone *J. Min. Sci.* 51 (6) 1063–1069
[7] Serdyukov SV Kurlenya MV Patutin AV Rybalkin LA Shilova TV 2016 Experimental Test of Directional Hydraulic Fracturing Technique *J. Min. Sci.* 52 (4)

Spatial discrete element analysis of problem on floor buckling in underground openings

SV Klishin* **and AF Revuzhenko****

Chinakal Institute of Mining, Siberian Brunch, Russian Academy of Science, Novosibirsk, Russia

E-mail: *sv.klishin@gmail.com, **revuzhenko@yandex.ru

Abstract. The authors analyze 3D problem on instability of a long arch cross-section opening using the method of discrete elements. Under analysis is loading by gravitational and tectonic stresses. Dilatancy, dry friction and viscosity are taken into account. It is assumed that contacting particles have an in-between elastic bonding element that fails under critical load. The article presents kinematic patterns of deformation and failure of surrounding rock mass around an underground opening.

1. Introduction

Buckling is one of the most widely spread processes induced by overburden pressure. Investigation of different types of buckling, review of the current knowledge on buckling mechanism and elaboration of measures aimed to predict and prevent such events is an important task, both theoretically and practically.

Consider a boundary value problem. Let an arch cross-section opening be driven in a rock mass at a depth H from free surface (Figure 1). The task is to define stress state of surrounding rock mass and the nature of buckling in the floor of the opening.

The in situ researches undertaken by different scientists show that buckling takes place in floor rocks when they are weaker than rocks in sidewalls and roofs of underground excavations. The explanation of the reason for the deformation reads that stronger rocks underlaid by weaker rocks act as a block die and deform considerably less as compared with the weaker underlying rocks.

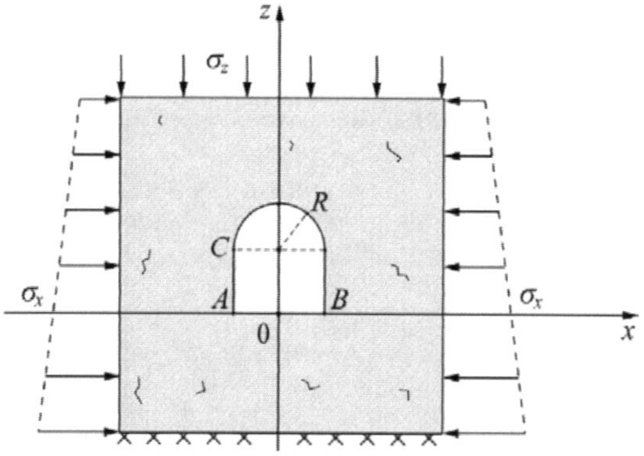

Figure 1. Schematic representation of the problem on floor buckling in underground opening.

The theoretical approaches to this problem can conventionally be divided into two groups. The first group are engineering solutions based on the hypothesis of a rigid block die [1–3]. These solutions use theories of limiting state and mostly consider deformation only of a part of rock mass (floor or roof). The material constraint of this approach is disregard of the rock mass loading history before limiting state. The second group rest upon the theory of elasticity and plasticity. The scope of stress state analysis in this case covers entire rock mass surrounding an underground opening [4–8]. Since displacements are small in elastic or elastoplastic formulations, the inference on buckling is based on the assumption of the floor rock loosening in the zone of rock crushing and on the analysis of strain field resultant from a boundary value problem solution.

The concept of rock mass as a medium with inner sources and sinks of energy backs the two-dimensional mathematical model constructed in [9, 10], considering internal block structure of rock mass, anisotropy of properties and sliding at block interfaces with regard to strength loss. The model is used to analyze numerically a problem on deformation of heavy rock mass surrounding a long horizontal opening with an arch cross-section. It is shown that zones of weakening and residual strength develop successively both in sidewalls and in floor and, finally, embrace the whole perimeter of the opening. The process of deformation is accompanied by uncontrollable dynamic release of elastic energy accumulated in the rock mass.

Evidently, the continuum formulations based on the methods of the elasticity and plasticity theory need no hypothesizing on limiting states. At the same time, it is required to construct closed systems of equations. At present, generally acknowledged equations of deformation of a medium remain yet to be developed.

Under such circumstances, it seems relevant to select an approach based on the discrete element method (DEM). This method allows determining evolution of stress state in rock mass both at the pre-limiting stage and at the stage of transition to localization of shearing and failure. The method is advantageous for needing no hypotheses on limiting state and no continuum equations to be postulated.

Originally, DEM was applied to solution of rock mechanics problems [15]. Now the method enjoys wide popularity in solving many problems in geomechanics [16, 17]. The essence of the method is replacement of a real medium by a packing of particles with introduction of laws of the particle interaction. The shape of the particles is a free parameter to be selected based on specific considerations. Mostly, it is assumed that the discrete elements have spherical form with the preset radii. This method induces no additional difficulties in solving problems with high (finite) values of strains and rotations. Thus, this method is a fundamental alternative to classical methods based on conventional knowledge on the continuum mechanics.

2. Discrete element method

According to DEM, a particulate material is represented by a set of N spherical particles (discrete elements). Each i-th element has a radius r_i and possesses the known physical properties: density, elastic and viscous moduli, friction, cohesion, etc. ($i=1,\ldots,N$). Movement of a discrete element includes translational and rotational motions, and is given by

$$m_i\ddot{\boldsymbol{x}}_i = \sum_j \boldsymbol{f}_{ij} + m_i\boldsymbol{g}, \qquad (1)$$

$$I_i\ddot{\boldsymbol{\theta}}_i = \sum_j (\boldsymbol{r}_{ic} \times \boldsymbol{f}_{ij} + \boldsymbol{M}_{r,ij}). \qquad (2)$$

Here, points mean differentiation with respect to time t; $\boldsymbol{x}_i$ is the radius-vector of the center of gravity of an i-th particle; $\boldsymbol{\theta}_i$ is the rotation of an i-th particle relative to the coordinate axes; m_i is the mass; I_i is the moment of inertia; $\boldsymbol{g}$ is the gravitational acceleration; $\boldsymbol{r}_{ic}$ is the vector directed from the center of an i-th particle to the point of contact; the force $\boldsymbol{f}_{ij}$ acts on an i-th particle from the side of a j-th particle and depends on the overlap of the particles and on the elastic and viscous moduli; $\boldsymbol{M}_{r,ij}$ is the moment of rolling resistance between particles or along the domain boundary. Summing in (1) and (2) involves all elements and boundaries that contact the current i-th particle. The shape of discrete elements is assumed unaltered for the period of contact; therefore, the degree of deformation of the elements is characterized by the value of overlap of contacting particles.

With the data on the law of motion and contact interaction between particles (equations (1) and (2)), as well as with the initial and boundary conditions, it is possible to construct solutions of the equations and to define evolution of stress state of a medium.

The program support designed based on DEM at the Institute of Mining, SB RAS enables numerical consideration of different modes of motion in a granular material composed of individual spherical particles [18, 19].

Below an illustration is given of the capacities of this method in the analysis of stress state in rocks and particulate media. Let a material specimen be composed of individual discrete particles (Figure 2a). The particles have the parameters: density $\rho_i = 2500$ kg/m^3, elasticity modulus $E_i = 3$ GPa, Poisson's ratio $\nu_i = 0.25$, external friction angle $\varphi_{ij} = 25°$, the diameters of the particles are selected from the uniform distribution over the interval from 0.0015 to 0.0025 m. The boundaries of domain of the specimen are perfectly smooth, the force of gravity is absent. Loading is implemented as uniform compression at a constant pressure $P = 10^4$ Pa and with a later on increase in σ_z oriented in parallel to the vertical axis towards the material. The overall deformation ε_z upon completion of loading was 0.3. Figures 2b and 2c show the plots of the relative stress σ_z / P and the relative volume V_r depending on the density of the initial packing. The curve 1 illustrates a dense packing, the curve 2 characterizes a loose packing. Obviously, it is possible to obtain materials with different strength properties by simply and only changing the initial density of specimens. The resultant consistent patterns are confirmed by in situ measurements and laboratory experiments [20].

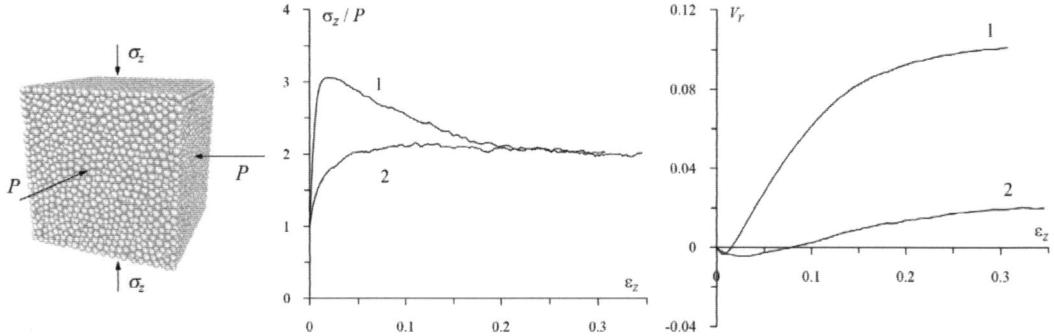

Figure 2. (a) The numerical experiment on triaxial compression of a particulate material; (b) stress–strain relationship in the tests; (c) change in specimen volume under compression: 1—dense packing; 2—loose packing.

A rock is a cohesive material; for this reason, the described model of particulate material deformation needs introduction of an additional bonding element (cohesion) between contacting particles. With this end in view, we use the procedure from [21–23].

The increments in the forces and moments induced in the bonding element by the particle interaction are given by

$$\begin{aligned} \Delta\hat{F}_n &= -\hat{k}_n A \Delta u_n, \ \Delta\hat{F}_t = -\hat{k}_t A \Delta u_t, \\ \Delta\hat{M}_n &= -\hat{k}_t J \Delta\theta_n, \ \Delta\hat{M}_t = -\hat{k}_n I \Delta\theta_t, \end{aligned} \tag{3}$$

where A is the intersection area between a particle and the bonding element; I and J are the inertia moment and the polar moment of inertia of the bonding element, respectively (Figure 3); Δu_n, Δu_t, $\Delta\theta_n$ and $\Delta\theta_t$ are the increments in the components of velocities and angular velocities in the normal and tangential directions, respectively.

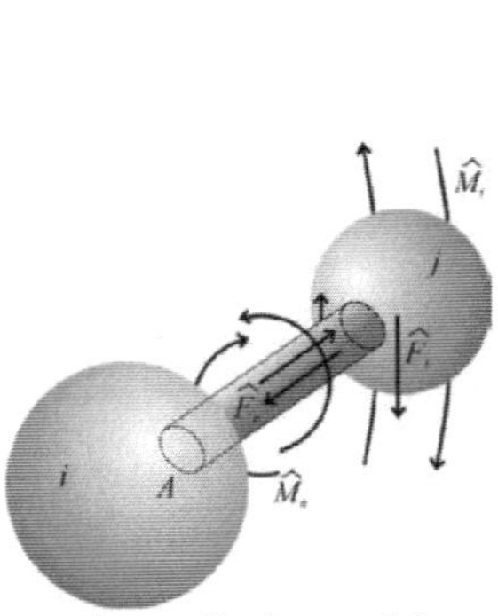

Figure 3. Schematics to find actual forces in bonding element.

The maximum tension force and shear force caused in the bonding element by relative motion of particles are

$$\hat{\sigma}_{max} = -\frac{\hat{F}_n}{A} + \frac{\hat{M}_t \hat{R}}{I},$$
$$\hat{\tau}_{max} = -\frac{\hat{F}_t}{A} + \frac{\hat{M}_n \hat{R}}{J}, \quad (4)$$

where $\hat{R} = \min(r_i, r_j)$. When the maximum tension force in the bonding element exceeds the tension bond strength ($\hat{\sigma}_{max} > \sigma_c$), or the maximum shear force is higher than the shear bond strength ($\hat{\tau}_{max} > \tau_c$) of the material, the bonding element fails and henceforth the particles are only subjected to the forces described by the equations (1) and (2).

By way of illustration, Figure 4 depicts a numerical experiment on destruction of a cylindrical specimen simulating a rock core composed of individual particles with cohesion under compression. It is seen in Fig. 4b that the load distribution line contains both elastic strain branch, and branches of weakening and residual strength typical of destruction processes in real materials.

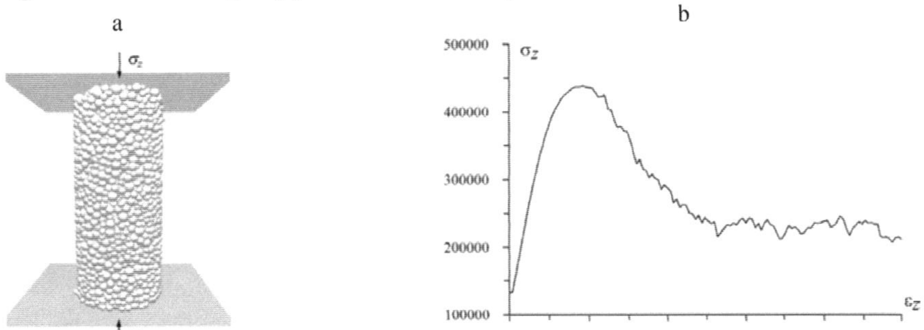

Figure 4. Uniaxial compression of rock specimen—numerical experiment: (a) schematic of the experiment; (b) stress–average strain curve.

In this way, by introduction of cohesion between discrete elements constituting a particulate material, it is possible to describe deformation and failure processes in cohesive rocks. Now we address again the problem on floor buckling in an underground opening. Figure 5 demonstrates the schematic of 3D numerical experiment. The experimental domain is represented by $N = 92000$ discrete elements with the radii selected from a uniform distribution over the interval from 0.05 to 0.15 m.

In the problem under discussion, the opening has a width of 3 m and a height of 3.5 m; the size of the domain in the Oy axis direction is 2 m. The physical parameters of the particles are: density $\rho_i = 2500$ kg/m^3, elasticity modulus $E_i = 8$ GPa; Poisson's ratio $v_i = 0.25$ and external contact friction angle $\varphi_{ij} = 28°$. The stiffness coefficients of the particle bonding in the normal and tangential directions: $\hat{k}_n = \hat{k}_t = 10$ MPa; the critical bonding strengths: $\sigma_c = 8$ MPa and $\tau_c = 5$ MPa. The boundaries of the domain are rigid and perfectly smooth walls. The gravitational vector is oriented vertically downward and its magnitude is 9.81 m/s^2. The lower boundary is assumed fixed.

The loading process involves two methods. The first method is moving the upper boundary at a constant low velocity u_z = const in the direction of the opening, with the left-hand and right-hand boundaries being fixed. The second method is shifting the left- and right-hand boundaries at a velocity u_x = const in the direction of the opening, with the upper boundary being fixed.

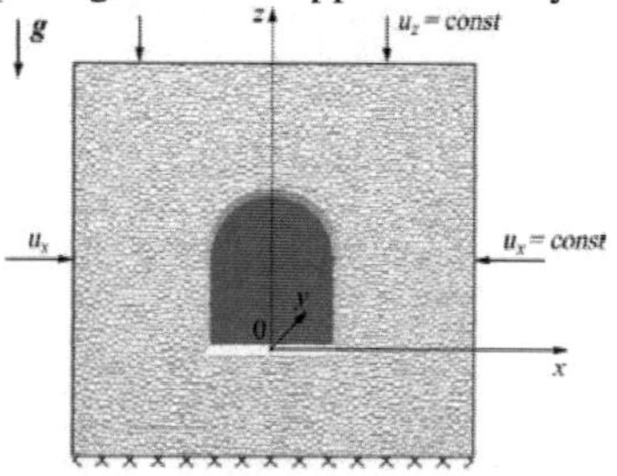

Figure 5. Numerical experiment on deformation of rock mass with a weakening in the form of an arch cross-section opening.

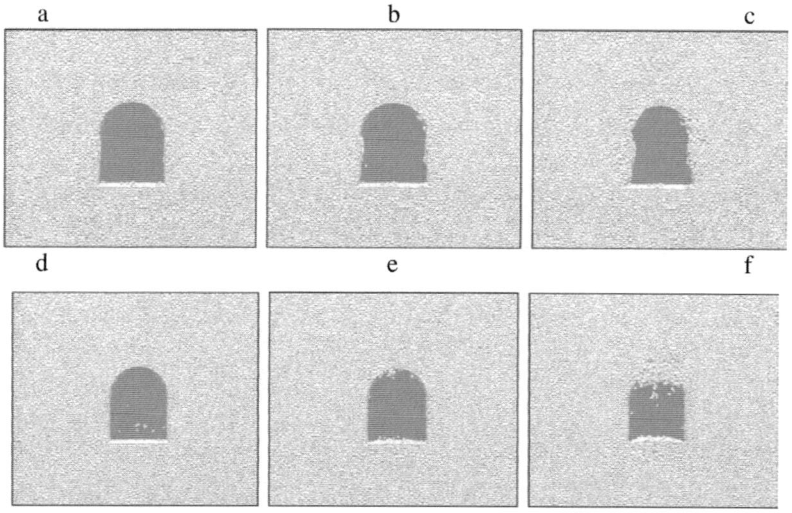

Figure 6. Deformation of surrounding rock mass around an opening at fixed moments of time: (a)–(c) action of gravitational stresses; (d)–(f) action of tectonic stresses.

The numerical experiment has discovered essential dependence of the opening perimeter deformation on the rock mass deformation history as a function of deformation method. When it is assumed that gravitational (vertical) stresses prevail, the opening sidewalls deform as Figure 7a–7c show. For dominating tectonic (horizontal) stresses, the failure pattern is shown in Figures 7d–7f where the instability and failure occur in the roof and floor of the opening.

3. Conclusion

To analyze buckling of floor rocks in mines, it seems most relevant to use the discrete element method. Under prevailing gravitational (vertical) stresses, failure takes places in the sidewalls of underground openings. Under dominating tectonic (horizontal) stresses, instability and failure occur in the floor and roof of the openings.

4. Acknowledgements

The investigation was done under support of Russian Science Foundation, project № 16-17-10121

IOP Conf. Series: Earth and Environmental Science **53** (2017) 012012 doi:10.1088/1755-1315/53/1/012012

References

[1] Tsimbarevich PM 1948 *Rock Mechanics* Moscow: Ugletekhizdat (in Russian)

[2] Chernyak IL 1978 *Prevention of Floor Buckling in Underground Excavations* Moscow: Nedra

[3] Sokolovsky VV 1960 *Statics of Granular Medium* Moscow: Fizmatlit (in Russian)

[4] Kovalenko VV and Ryazantsev AP 2013 *Evaluation of Process Variables for Prevention of Floor Rock Buckling in Coal Mines* Dnepropetrovsk: Nats. Gorn. Univ. (in Russian)

[5] Voloshin VA, Rib SV, Fryanov VN and Cherepov AA 2015 Formation of increased overburden pressure zones under influence of coal pillar-block die in bed series mining *GIAB* No 7 pp. 23–29

[6] Khomyakova A.A., Nikitina A.M., Rib S.V., Borzykh D.M. Analysis of influence exerted by a diminishing coal pillar on floor rock buckling in a roadway. In: High Technologies of Mineral Mining and Utilization: Collection of Papers. V.N. Fryanov (Ed.). Novokuznetsk: Sib. Gos. Industr. Univ., 2014, pp. 118–123. (in Russian)

[7] Demin VF, Yavorsky VV, Demina TV and Steflyuk YuYu 2015 Efficient methods to prevent floor rock buckling in gateways in coal mines *Usp. Sovr. Est. Nauk* No 12 pp. 95–99

[8] Kuzmin SV and Salvasser IA 2014 Factors and classification criteria of buckling *Byull. KuzGTU* No 3(103) pp. 43–44

[9] Lavrikov SV, Mikenina OA, Revuzhenko AF and Shemyakin EI 2008 The concept on non-Archimedean multiscale space and models of plastic media with structure *Physical Mesomechanics* Vol 11 N 3 pp. 233–246

[10] Lavrikov SV 2010 Calculation of stress state in weakening block rock mass around an opening *Physical Mesomechanics* Vol. 13 No 4 pp. 53–63

[11] Labuz Joseph F and Zang A 2012 Mohr–Coulomb failure criterion *Rock Mechanics and Rock Engineering* Vol. 45 No 6 pp. 975–979

[12] Abbo AJ and Sloan SW 1995 A smooth hyperbolic approximation to the Mohr-Coulomb yield criterion *Computers and Structures* Vol. 54 No 3 pp. 427–441

[13] Hoek E and Brown ET Empirical strength criterion for rock masses *J. Geotech. Engng. Div., ASCE 106(GT9)* pp. 1013–1035

[14] Hoek E 1983 Strength of jointed rock masses *Géotechnique* Vol 23 No 3 pp. 187–223

[15] Cundall PA and Strack ODL 1979 A distinct element model for granular assemblies *Géotechnique* Vol 29 No 1 pp. 47–65

[16] Klishin SV, Klishin VI and Opruk GYu 2013 Modeling coal discharge in mechanized thick and steep oal mining *Journal of Mining Science* Vol 49 No 6 pp. 932–940

[17] Lavrikov SV and Revuzhenko AF 2015 DEM code-based modeling of energy accumulation and release in structurally heterogeneous rock masses *AIP Conference Proceedings 1683*

[18] Klishin SV and Revuzhenko AF 2014 3D discrete element approach to Janssen's problem *Journal of Mining Science* Vol 50 No 3 pp. 417–422

[19] Klishin SV and Revuzhenko AF 2015 A class of vortex flows in granular medium *Journal of Mining Science* Vol 51 No 6 pp. 1070–1076

[20] Bishop AW 1972 Shear strength parameters for undisturbed and remoulded soil specimens *Stress–Strain Behavior of Soils: Proceedings of the Roscoe Memorial Symposium* Parry RHG (Ed) Cambridge University pp. 3–58.

[21] Potyondy DO and Cundall PA 2004 A bonded-particle model for rock *International Journal of Rock Mechanics and Mining Sciences* Vol 41 No 8 pp. 1329–1364

[22] Mohammad Sadegh Asadi, Vamegh Rasouli and Giovanni Barla 2012 A bonded particle model simulation of shear strength and asperity degradation for rough rock fractures *Rock Mechanics and Rock Engineering* Vol 45 no 5 pp. 649–675

[23] Obermayr M, Dressler K, Vrettos C and Eberhard P 2013 A bonded-particle model for cemented sand *Computers and Geotechniques* Vol 49 pp. 299–313

Evaluation of a new trend in the geotechnology for steeply dipping vein deposits in Kyrgyzstan

KCh Kozhogulov[1] **and ZhM Ganiev**[2]

[1]Institute of Geomechanics and Mineral Mining, National Academy of Sciences of the Kyrgyz Republic, Bishkek, Kyrgyz Republic
[2]Institute of Mining and Mining Technologies, Bishkek, Kyrgyz Republic

E-mail: [1]ifmgp@yandex.ru, [2]igdigt@inbox.ru

Abstract. The authors evaluate advancement in the geotechnology for steeply dipping vein deposits. It is proposed to use large-diameter drilling with drilling assembly BGA-2V-02. As a result, the efficiency of mining is enhanced.

Lodes or vein deposits are important sources of minerals in the world. In the USSR in the 1970–80s, underground mining at such deposits yielded round the fourth of the total production of nonferrous, rare and precious metals in the country. Almost 70 mines performed development of such deposits under different ground conditions [1]. Regarding the world, such deposits are widely spread, e.g. in the North America.

Vein deposits are developed in the world using open stoping, systems with caving and systems with artificial support of mined-out voids. In accordance with [1], steeply dipping ore body development with vein mining method makes 55–60%, cut-and-fill stoping—20–25% and sublevel stoping—8–10%.

Performance of such mines greatly depends on the thickness and inclination of veins. Cut-and-fill methods show higher figures, while vein mining and sublevel stoping have lower performance.

Vein deposits attract increasingly much interest in the Kyrgyz Republic in the latest years. These are such ore bodies as Ishtamberdy, Zhamgyr, Tereksai and Terekkan. The other vein deposits—Kurandzhailyau, Altynzhylga, Komator, Savoyardy, Togolok, Pervenets, Trudovoe—are planned to be put into operation in the short term.

A feature of these veins is their steep dipping and low strength. The dip angles of these veins range from 52 to 85 deg and thickness varies from 0.3 to 1.5 m.

Steeply dipping veins are mined in Kyrgyzstan using systems with ore shrinkage (Zhamgyr, Terekkan, Trudovoe), with systems with support and selective extraction (Kurandzhailyau) and with sublevel stoping (Ishtamberdy). The loss of ore varies from 6.5% (Terekkan) to 14.8% (Ishtamberdy) while dilution reaches 30.7% (Ishtamberdy).

The analyses of ore loss and dilution at some vein ore bodies (Russia, Kazakhstan, Uzbekistan) show wide scatter of figures, depending on ore body occurrence and mining technology. In Kyrgyzstan, conventional mining of veins is connected with high dilution (to 31.0%) and loss (to 13%), which entails high operating cost.

For this reason, the Institute of Geomechanics and Mineral Mining, National Academy of Sciences of the Kyrgyz Republic jointly with the Institute of Mining and Mining Technologies carry out research aimed to justify a new trend of geotechnology for vein deposits. As a result of the research, an underground geotechnology has been substantiated for low-strength ore veins and includes large-diameter drilling. This geotechnology is based on rotary drilling employed in coal mines. Rotary drilling of coal was widely used in the world [3] and in thin coalbed mining in the USSR [4–7].

Recent models of rotary drilling machines in USA were of high technical excellence and ensured sufficient technical-and-economic indices of face operation and, first of all, labor production output.

On the other hand, the high performance was reached in mining of thick beds of weak coal. Later on, rotary drilling machines were less and less employed as they failed to ensure high mining rates as compared with the continuous heading machines. In the USSR rotary drilling was applied mainly to extraction of thin (0.5–0.9m) beds. Rotary drilling machines and plants were designed and manufactured for those ground conditions. Russia preserves the same approach to rotary drilling machines, it is noteworthy that coal mining in Russia, to be profitable, presumes low cost of final product at comparatively low mine capacity—in small mines with thin personnel operating with high production output. In such conditions, rotary drilling machines are the best choice for their operation involves [7]:

—minimum number of personnel (3–4 miners per shift);

—much lower cost of equipment as compared with the conventional technologies;

—drop in operating cost of mining (2–3 times) owing to reduced consumption of materials (explosives, emulsion, spare parts, etc.);

—enhanced safety and comfort of operation.

The idea of using rotary drilling method in vein mining in Kyrgyzstan was first trialed at Ishtamberdy deposit. The equipment was drilling rig BGA-2V-02 adjusted for the local conditions.

Drilling rig BGA 2V is meant for hole drilling and expansion with cutter bits in rocks with a strength up to 50 MPa and with rotary bits in rocks with a strength up to 80 MPa. Drilling if performed in underground excavations with a cross section not less than 4 m^2. The minimum thickness of a vein is to be 0.3 m. Drilling rig BGA 2V is equipped with a rotary drill.

The rig (Figure 1) is composed of a drilling assembly 1 and an oil station 2 with a radiator 3. The oil station is mounted with screws on the frame of the drilling assembly. The rig performs all drilling operations.

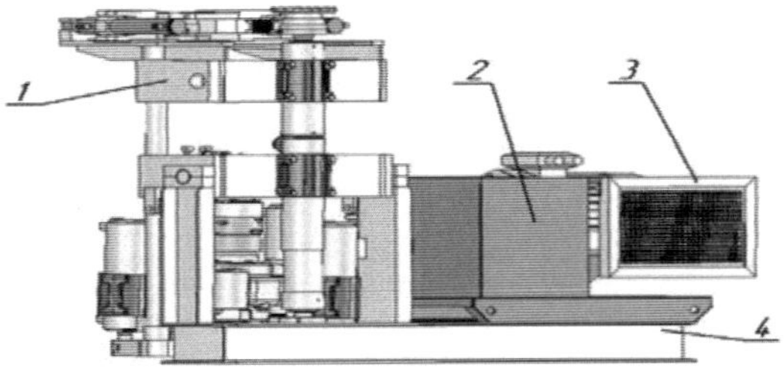

Figure 1. Drilling rig BGA 2V: 1—drilling assembly; 2—oil station; 3—radiator; 4—frame.

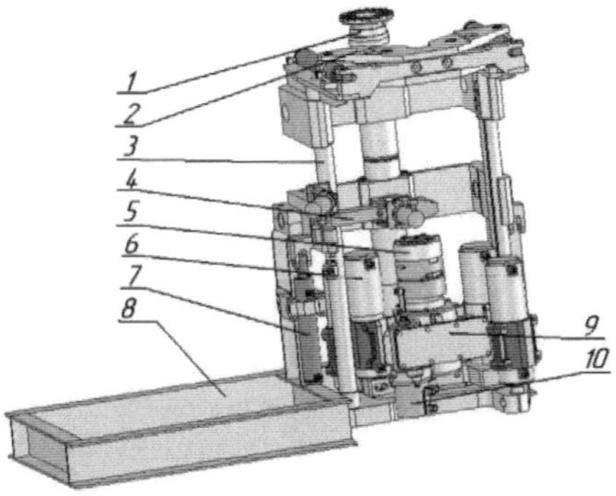

Figure 2. Drilling assembly: 1—leg; 2—crown runner; 3—rotating frame; 4—manipulator mechanism; 5—tool-joint; 7—rotation cylinder; 8—frame; 9—reducing gear; 10—fluid-power motor.

The drilling assembly (Figure 2) contains reducing gear 9, cylinder 6 and 7, joint 5, crown runner 2, rotating frame 3, frame 2 and manipulator mechanism 4. The rod of the cylinder 7 is connected with the rotating frame 3, and the body of the cylinder is mounted on the frame 8. The crown runner is screwed to the rotating frame. The bodies of the cylinders are mounted on the reducing gear, and their rods are hinged with a yoke. The manipulator mechanism has a plug valve to reduce pressure and block the feed.

Rotation is executed by fluid-power motors connected with gear shafts by gear wheels. The gear shaft transmits rotation via gear wheels to a shaft where the tool-joint is set.

The oil station converts electric power from the power grid into hydraulic energy. The electric motor, via a tooth clutch, transmits rotation to a two-section gear-type pump. The larger section of the pump is for rotation of the fluid-power motors and for maneuvering feed. The smaller section of the pump is for setting the rig at a certain angle, to feed the hydraulic leg, control the crown runner and manipulator mechanism and for regular feed. The oil station is composed of a tank, pump drive and radiator. The radiator is equipped with a fan rotated by the fluid-power motor sequentially connected with a drain line.

The drilling assembly has a dust suppression system with water feed from a mine water line. The assembly is run by 2 operators.

The proof tests of the drilling assembly BGA-2A were carried out on 7–11 July, 2014 on level 2216 in drift no. 2 on the west of the northern ore vein. Total length of drilling was 30 running meters of holes drilled from level 2216 to level 2244.6; 26.5 running meters of holes were drilled with $D = 300$ m, $d = 500$ m, and 3.5 running meters of holes were drilled at $d = 150$ m and at an angle of 62 deg. Based on the test results, an expert committee decided on continuing the tests of a retreat reamer from level 2244.6 and proved applicability of the drilling assembly BGA2V in low-strength thin ore veins.

References

[1] Kursakin GA 2002 *Gold Vein Mining Tehnology* Vladivostok: Dalnauka (in Russian)
[2] Mining activity in the Western world *Mining Mag.* 1987 Vol 160 No 1 pp 41–52
[3] Kryuchkov VV 1963 *Rotary Drilling in Coal in Open Pit Mines in the U.S.A.* Mosow (in Russian)
[4] Sudoplatov AP and Ivanov KI 1962 *New Highly Productive Methods of Coal Extraction* Moscow (in Russian)
[5] Grinko NK 1976 Trends in unmanned coal mining *Ugol* No 6
[6] Titov NV, Kochergin VI, Turuk YuV and Kurnakov VA 2006 Prospects for rotary drilling in coal mines of the Russian Donbass *Izv. Vuzov Sev.-Kavkaz. Reg.* pp 56–59
[7] Galchenko YuP and Sabyanin GV 2011 *Geotechnology for Vein Deposits* Nauchtekhlitizdat (in Russian)

Advancement in modern approaches to mineral production quality control

EV Freidina[1,2], AA Botvinnik[1] and AN Dvornikova[1]

[1]Chinakal Institute of Mining, Siberian Branch, Russian Academy of Sciences, Novosibirsk, Russia
[2]Novosibirsk State University of Economics and Management, Novosibirsk, Russia

E-mail: [1]albyna@misd.ru

Abstract. The natural resource potential of mineral deposits is represented by three categories: upside, attainable and investment. A modern methodology is proposed in this paper for production quality control, and its tools aimed at ensuring agreement between the product quality and the market requirements are described. The definitions of the costs of the product quality compliance and incompliance with the consumer requirements are introduced; the latter is suggested to use in evaluating resource potential of mineral deposits at a certain degree of probability.

A mineral deposit as a raw material object has a definite resource potential. The resource potential is understood as the cumulative natural reserves that are or can be used in economic life given the actual technical capabilities and social and economic reality. Capabilities of a society is a variable that defines stages of progressive development in technology of mineral mining and processing. From that standpoint, the authors put forward an approach to evaluation of a resource potential classified into upside, attainable and investment potentials [1–3].

Upside potential is defined as the maximum value of mineral reserves consummated using technologies that ensure complete extraction a mineral and preservation of its natural quality.

Attainable potential is a value of mineral consummated given the key mineral mining and processing technologies now in use, i.e. this is a strategic resource of the industry.

Investment potential is an ability of a mine to produce and supply the market with a merchantable product ensuring acceptable profit under efficient use of all required natural and technological reserves.

The resource capacity of the upside, attainable and investment potentials is a frozen in time value. The size of the upside potential is a forecast of mining capabilities of a mineral deposit and always entails a certain risk associated with confirmation of proven reserves. The attainable potential is even more variable in time as the current and new technologies of mineral mining and processing undergo continuous improvement and development. The attainable potential grows with the diversification of different quality mineral products. For instance, coal products are marketable in fuel engineering (energy sector, coke chemistry) and in other industries (chemistry, pharmaceutics, agriculture, food industry, road construction, etc.).

Variation of the upside and attainable potentials changes the investment potential of a mineral deposit. The investment potential is a conversion of an attainable potential into a marketable product under prevailing conditions and considering quality requirements imposed by a consumer, market situation and prices. Adaptational capacity of a mineral producer should be flexible and efficient. Such adaptability has a few grounds.

First, the detailed and operational exploration of a mineral deposit under mining, as a rule, introduces modification in the estimates of ground conditions, mineral reserves and quality. It becomes necessary to have a selection of options for expansion or reduction in the range of different quality products with a view to deciding on an optimal alternative under prevailing conditions, considering mineral reserves and market demand.

Second, it is not impossible that demand of actual and potential consumers for a certain composition and quality product will change. Moreover, demands of consumers are unevenly distributed in time. Based on high variability and uncertainty of consumer behavior, it is required that the investment potential of a mineral deposit can be promptly re-evaluated so that to optimize interaction between a producer and a consumer under prevailing conditions.

Third, the investment potential is attackable by unpredictable and mine-independent fluctuations in the sphere of economics and finance.

Reduction in the uncertainty of investment potential of a mineral deposit is possible with the state-of-the-art procedure of product quality planning such as Quality Function Deployment (QFD) [4]. QFD method is an ordered way of transforming demands or requests of a consumer into quality standards for an anticipated product. QFD is applicable to planning any production, processes and operations aimed to improve product quality and competitive recovery.

Regarding mineral product quality planning, the basic procedures and instruments of QFD are listed below.

1. Determination of wants and detailed requirements of a consumer per key indexes of marketable product quality, rating of criticality of each requirement and allowable range of indexes.

2. Assessment of consistency between the key indexes of mineral quality with the product quality standards imposed by a consumer, setting of mineral quality targets and evaluation of quality indexes to be improved in order to satisfy the demand for a marketable product.

3. Determination of best capabilities (technologies) of a producer to achieve product quality standards to extremely satisfy consumer demands. Evaluation of technical difficulties of achievement of mineral quality targets.

4. Valuation of meeting consumer demands for mineral quality.

5. Estimate of competitors in satisfaction of consumer demands.

6. Evaluation of quality control methods that ensure efficiency of production and commercial activities in mineral mining.

7. Formulation of objectives and development of strategy of mineral quality control.

A worthy technique of QFD is the method of Product Planning Matrix capable to integrate an information field composed of (Figure 1):

—natural quality of a mineral;

—quality indexes to be improved to meet quality standards set by a consumer;

—info about competitors: whether they better or worse fulfill consumer demands;

—capability of a mine to outrun competitors;

—investment required to reach an agreement between marketable product quality and quality standards of a consumer.

Each step of the Product Planning Matrix is a sub-result of research aimed at the adequately supported representation of local matrices:

—matrix of relationship between consumer demands and producer capabilities;

—correlation matrix of links between adjustable quality indexes;

—matrix of evaluation of competitor supremacy (consumer benchmarking);

—matrix of evaluation of technological supremacy of competitors in reaching product quality targets (technical benchmarking);

—matrix of assessment of technical difficulties of reaching product quality targets;

—matrix of value of mineral mining at reached product quality targets.

An illustration of the matrix of meeting consumer demands for coal product quality is shown in Figure 2.

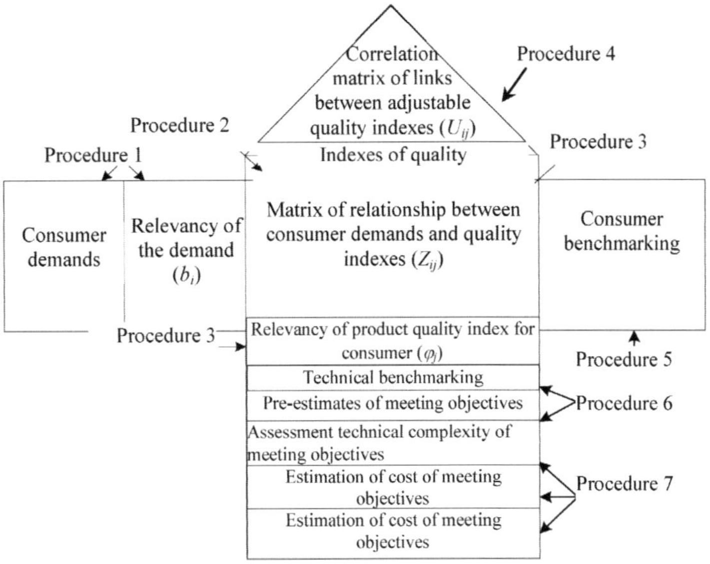

Figure 1. Product Planning Matrix.

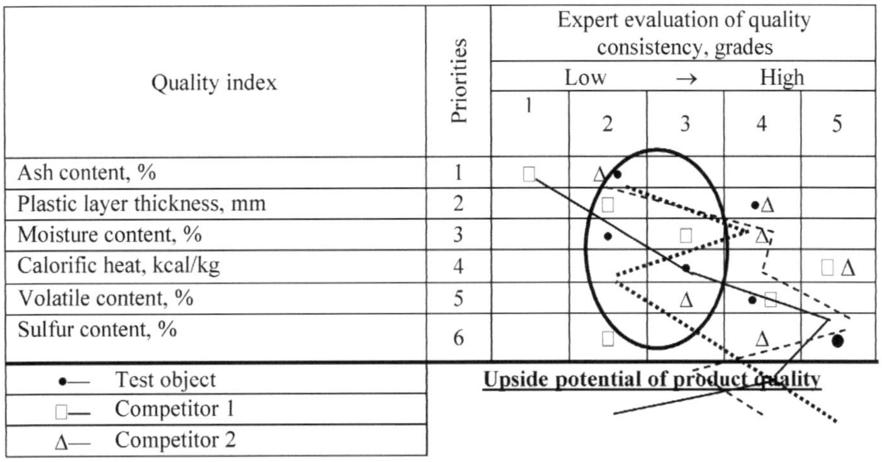

Figure 2. Matrix of evaluation of competitors.

Development of QFD in short-range and long-range planning of mineral quality will require that mines prepare a package of documents on systemic management of product quality control. The framework of the QFD package software systems, procedures, methods, quality maps and some other intelligent products [4] mostly represented by trial specimens that successively pass approval finalization of which needs investment from operating mines.

Influence of mineral quality on efficiency of mining investment potential shows on supply of mineral products to consumers. Relationship between a mine–producer and an energy object–consumer of coal products has a feature: there is an estimated value of violation of quality standard set in a contract (price correction factor for fuel). For this reason, the issues of cost of coal quality conformity/nonconformity with consumer standards are to be handled at the stage of mining.

The nonconformity cost is exemplified in terms of the Novosibirsk Heat Power Plan. The price correction factors for violation of standards set for basic adjustable indexes of fuel quality are assumed as

—ash content: 1.95% discount per each percent of increment in mass fraction of ash content of ROM coal;

—moisture content: 1.69% discount per each percent of increment in mass fraction of moisture content of ROM graded coal.

In this connection, it is suggested to consider two measures of evaluation for each λ-th degree of quality (coal grade): conformance of an i-th quality index with consumer standards ($\varphi_{\lambda i}$) and cost accordance CA_λ [5].

Quality conformance of a mineral product is achieved when an average value of an i-th quality index ($\overline{X}_{\lambda i}$) is within the allowable range ($\Delta_{\lambda i}$) of a consumer-set standard $X^{н}_{\lambda i}$

$$X^{н}_{\lambda i} - \Delta_{\lambda i} \leq \overline{X}_{\lambda i} \leq X^{н}_{\lambda i} + \Delta_{\lambda i}. \quad (1)$$

When quality indexes that worsen final product quality, e.g. contents of ash, moisture, sulfur and volatiles for coal, exceed the allowable range, the nonconformity is given by

$$X_{\lambda i} - (X^{н}_{\lambda i} + \Delta_{\lambda i}) = \Delta X_{\lambda i}, \quad (2)$$

and the degree of nonconformity is

$$\omega_{\lambda i} = [\overline{X}_{\lambda i} - (X^{н}_{\lambda i} + \Delta_{\lambda i})] / (X^{н}_{\lambda i} + \Delta_{\lambda i}). \quad (3)$$

In this case, the degree of conformity

$$\varphi_{\lambda i} = 1 - \omega_{\lambda i}. \quad (4)$$

Inasmuch as violation of quality indexes of a delivered product as against the contractual product standards is estimated in the same units of measure (percentage) and the cost of nonconformity of a negative quality index is evaluated in units of money, it is possible to reduce the oval nonaccordance cost to a generalized index given by

$$CN_\lambda = \sum_{i=1}^{M} \Delta X_{\lambda i} r_{\lambda i}, \quad (5)$$

where M—number of quality indexes for the given subject of power engineering; r_i—rate for overrun of allowable value of a negative quality index.

The anticipated returns from sales or the cost of the final product meeting the consumer quality standards is determined as

$$C = \sum_{\lambda=1}^{N} V_\lambda \gamma_\lambda p_\lambda, \quad (6)$$

where N —number of degrees of quality of coal grades; V_λ, γ_λ, p_λ—production output, yield of concentrate of a λ-th quality degree or grade of coal and the concentrate price, respectively.

Let a power engineering plant is supplied with coal of the same quality or grade. Per each quality index, the degree ($\omega_{\lambda i}$) and the overall cost (CN_λ) of quality nonaccordance are determined. With a view to CN_λ, the accordance cost CA_λ will be evaluated as the difference between the anticipated returns and rate for unrated quality violations

$$CA_\lambda = V_\lambda \gamma_\lambda p_\lambda - CN_\lambda. \quad (7)$$

It follows from (7) that the actual returns of coal product sales will be

$$CA = \sum_{\lambda=1}^{N} (V_\lambda \gamma_\lambda p_\lambda - CN_\lambda). \qquad (8)$$

The quality nonconformity shows to a high probability when a consumer is supplied with ROM coal. The economic efficiency of the investment potential is to be reduced by the value of nonaccordance cost of the supplied product quality and the consumer-set quality standards.

Evaluation of mineral quality based on positive indexes should take into account that the product value is higher when a positive quality index is higher. For example, for black coal, the positive quality indexes are, primarily, calorific heat and plastic layer thickness. From the standpoint of a consumer, the degree of conformity between the supplied product quality and the consumer demand is given by

$$\omega_{\lambda i} = [\bar{X}_{\lambda i} - (X_{\lambda i}^{\text{н}} - \Delta_{\lambda i})]/(X_{\lambda i}^{\text{н}} - \Delta_{\lambda i}). \qquad (9)$$

So, a producer should control coal quality so that the coal quality indexes are always within the pre-set allowable ranges. Otherwise, coal price suffers considerable changes. The afore studies are based on an approach to enhancement of mining efficiency implemented in Neryungri Coal Field based on the recommendations made by the Eastern Research Institute for Coal Chemistry. It was advised to grade quality of local coking coal reserves based on plastic layer thickness and to set essentially different prices for each quality grade. In this case, the producer objective is to minimize quality divergence of a marketable product with respect to positive quality indexes, or the aggregate value of a coal product will drop and, accordingly, the economic efficiency of mining investment potential will decline.

Conclusion

Identification of categories of a mineral resource potential (upside, attainable and investment) enables elaboration of the perception of usability of mineral resources, their conversion into marketable product and efficient marketing of the latter.

The authors have proposed to add the value of an investment potential with the cost of nonconformity between the supplied coal product quality and the quality standards set by consumers—power engineering objects. It is found that when the supplied product is ROM coal, the quality nonaccordance is highly probable.

The investment potential sustainability is supported by the use of QFD method of product quality planning this method means a systemic transformation of the demands and wishes of a consumer into the measures of quality of the anticipated product. On the whole, the effectiveness of activities toward the development of the investment potential of mineral mining gets enhanced. Adherence to QFD method will allow a producer and a consumer to be sure that the planned quality of coal product will be attained and the producer will enjoy stable and successful position in the fuel market.

References

[1] Freidina EV, Botvinnik AA and Dvornikova AN 2005 Information technologies for resource potential control at a coal deposit *Surface and Undersea Mineral Mining: VIII International Conference Proceedings* Sofia

[2] Petrov OV (Ed) 2001 *Geological Dictionary* Saint-Petersburg: VSEGEI (in Russian)

[3] Fomin PA and Starovoitov MK *Features of Production Potential Evaluation* Available at: http://www.cfin.ru/management/manufact/manufact_potential.shtml

[4] Freidina EV, Botvinnik AA and Dvornikova AN 2008 Coal quality control in the context of international standards ISO 9000-2000 *Journal of Mining Science* Vol 44 No 6 pp 585–599

[5] Freidina EV, Botvinnik AA and Dvornikova AN 2009 Selection of a criterion of conformance between coal product quality and consumer requirements *Gorn. Inform.–Analit. Byull.* No 5 (Special Issue) pp 90–101 (in Russian)

Geotechnical approaches to coal ash content control in mining of complex structure deposits

SA Batugin[*], VL Gavrilov[**] and EA Khoyutanov[***]

Chersky Institute of Mining of the North, Siberian Branch, Russian Academy of Sciences, Yakutsk, Russia

E-mail: [*]batuginan@mail.ru, [**]gvlugorsk@mail.ru, [***]evgeni_h@nm.ru

Abstract. Coal deposits having complex structure and nonuniform quality coal reserves require improved processes of production quality control. The paper proposes a method to present coal ash content as components of natural and technological dilution. It is chosen to carry out studies on the western site of Elginsk coal deposit, composed of four coal beds of complex structure. The reported estimates of coal ash content in the beds with respect to five components point at the need to account for such data in confirmation exploration, mine planning and actual mining. Basic means of analysis and control of overall ash content and its components are discussed.

The Institute of Mining of the North, SB RAS, has proposed a new approach to estimating ore dilution based on continuous expansion and upgrading of the knowledge and understanding of the nature of a geological environment [1], considering mechanisms of technological conversion of a georesource into a finished product [2]. This has enabled enhanced extraction of mineral reserves with the improved efficiency of mining based on the rational ore pretreatment.

The research is continued in terms of coal reserves. The present-day situation forces to launch mining operations in coal fields with the increasingly complex structure of beds, and, by analogy with ore deposits, the new knowledge is required to maximize the effect of coal quality management based on the available and new synthesized information on the georesource. With this end in view, it has been proposed to assess coal ash content as a composition of natural and technology-induced impoverishment [3, 4]. Owing to more comprehensive analysis of structure and occurrence of a mineral substance in coal, this approach allows finding potential ways of auxiliary preparation of coal for dressing on-site, in an open pit mine, based on generation of coal flows with more uniform process properties for individual preparation of these flows for the purposes of enhancement of dressing efficiency. Ash content is proposed to be presented as a sum of at least five constituents: (1) dirt in roof, floor and damage zones ($A^d{}_1$); (2) extractable dirt bands and high-ash coal bands inside a bed ($A^d{}_2$); (3) unextractable thin dirt bands ($A^d{}_3$); (4) mineral particles of sand, clay and dust impregnated in coal and filling joints and pores ($A^d{}_4$); (5) matrix inorganic mineral admixtures ($A^d{}_5$).

As a test object to estimate the constituents of ash content and to study controllability of this quality index, the western site of Elginsk deposit has been selected; this site is composed of four beds of a complex structure, which is confirmed and refined by analyses of geometrical parameters of these coal beds and spatial variability of ash content in them [5]. For such high-ash and difficult-to-process coals, it is of high concern to find ways of enhancing utilization efficiency. The main coal beds are represented by two flat-dipping series 5–10 m thick. The design method is run-of-mine output, and coal, being highly mineralized per se, is additionally impoverished due to numerous intra-bed dirt bands (more than 0.05 m thick) and thinner (less than 0.05 m) intercalations and lenses neglected by geologists in the reserves appraisal.

Currently the deposit is under complementary exploration, advanced sampling of coal and first-stage mining in the site chosen as a test object. In 2015 Elgaugol company produced 3.95 Mt of coal [6]. The preparation plant of the company produces concentrate to manufacture standard charred coal and to be used in PCI technologies. The structure of the preparation products, as compared with the products of the other preparation plants, e.g. in Neryungri, is ineffective due to low yield of the concentrate and high yield of tailings.

On the test site, the above listed constituents of the ash content index were assessed using the available and permanently updated Database of Elginsk Coal Sampling. The studies involved various geological information systems, and the known and new-developed and adapted calculation methods [5]. As a result, in the four main beds (with the total ash content from 26 to 40%), A^d_1 makes 4.6–7.5%; A^d_2—7.2–21%; A^d_3—4.9–8.4%; A^d_4—55.5–77.4%; A^d_5—2.0–10.2%. Mapping of ash content contours for individual constituents with a view to studying their spatial variability shows that the structure and combinations of the constituents in areas with the same values of total ash content differ. This means that an area of coal beds with the uniform values of the total ash content can be additionally ranged into zones where this or that constituent prevails. As a consequence, though potential complication of mine planning processes, we have a potential to arrange quality control over coal forwarded to preparation.

Development and implementation of the ash content control with regard to effect of individual constituents should be based on assuming the deposit–open-pit mine–preparation plant chain as a unified process space where various processes are integrated for enhancement of end-to-end efficiency, such as: (1) operating exploration and advanced sampling of a deposit; (2) long-term and current mine planning; (3) coal extraction in face areas of different geometry and production output; (4) preparation of coal to dressing by technological and organizational influence of generated coal flows (separation and fusion of coal flows, use of ground-type or "wheeled" storages, etc.).

Coal flows appear when coal mass turns to a broken mass of coal and rocks after drilling and blasting or re-excavation, i.e. when georesource passes from static state to dynamic state connected with the motion of excavated or prepared coal. The existing mining technology at Elgaugol provides for differentiation, multiple separation and fusion of produced coal flows. In this case, in focus is, first of all, oxidation of coal (characterized by the plastic layer thickness, free bloating index, and American standard oxidation index) and ash content (concentrate produced by preparation of borehole and face samples under laboratory conditions at the partition density of 1350–1400 kg/m^3, overall for a bed, regardless dirt bands thicker than 0.05 m, the so-called "net").

Inclusion of the ash content constituents in the coal product quality control can and should result in generation of new coal flows. The expediency of such differentiation ensues from the fact that processing of coal mass with more uniform quantitative characteristics, physical properties and grain-size composition will produce additional concentrate of higher value and marketability as compared with middlings. According to [7], simple reduction in size of Elginsk coal from -50 mm to -8 (-6) mm by re-milling to obtained better dissociation of highly mineralized coal and rock aggregates allow increment in the yield of the concentrate by 8–17% with the ash content of up to 10% at the concurrent reduction in the yield of middlings that are poorly merchantable. Based on the discussion and analyses of such notions as "limit potential" (maximum theoretically achievable value of mineral reserves) and "attainable potential" (recoverable value achievable using advanced technologies) in terms of Talda deposit in Kuzbass, it is clear that high technologies of mineral mining and processing as against the conventional run-of-mine output with overall coal preparation offer an increased potential of mine projects [8]. Application of the technology of thin blast-free slicing with Wirtgen surface miners at one of the Australian deposits, as compared with the common techniques, enabled high yield of the concentrate from 63 to 75% at the simultaneous decrease in the coal as content by 4% and the increase capacity of preparation plant [9].

In the conditions when it is necessary to enhance efficiency of utilization of coal reserves in each bed and each extraction area, a better insight into the nature and generating mechanisms of coal ash content assists in adaptation of these processes and/or their elements (in planning or in operation) to the changed understanding of georesources. The processes (1)–(4) specified above in this paper are capable of including any analysis tools and control methods to deal with the overall ash content and its constituents. Engineering decisions, technologies and process flow charts, or organization, informational and economic methods to generate and control coal flows are the result of analyses of a sufficient body of information, patent search and approaches developed with the participation of the present article authors. They have no pretensions to omnitude but they are fruitfully used in mines in Russian and/or in the world and are proposed in scientific literature and design documentation.

At the stages of supplementary study of georesources and mine planning, it is possible to use various analysis approaches or their combinations in order to enrich the knowledge on a study object and to generate rational plans of mining (Table 1: gray color marks the analysis tools meant either for an ash content constituent or for the overall ash content). Individually, some decisions and methods in regard to complex-structure coal deposits are discussed, for instance, in [10–16].

Table 1. Analysis tools useful at stages of supplementary exploration and mine planning.

Supplementary exploration and mine planning	A^d_1	A^d_2	A^d_3	$A^d_{overall}$
Studies and mapping of damage zones and interfaces of roof and floor rocks and coal				
Detection of geologically and geotechnically uniform extraction zones and their mapping per beds and extraction sites				
Estimation of potential reduction in ash content and its constituents (natural impoverishment)				
Mapping of contours of overall thickness of beds				
Analysis of ash contents in boundary layers (floor–roof of beds)				
Analysis of structural complexity, spatial variability of ash content in beds, evaluation of reliability				
Additional analysis of physical properties of coal and intra-bed intercalations				
Improvement of exploration methods and means and supplementary study of the constituents of ash content				
Express-methods of ash content study in coal–rock mass, analysis of petrography of coal substance				
Rating of coal reserves in regard to their preparation to mining, considering spatial variability of coal quality				
Planning of concurrent operation of production faces, accumulating and/or blending coal storage yards				

The commonly used tools of analysis (Table 1) acquire specific value and find new application in the context of the proposed differentiation. For instance, mapping of uniform zones of the overall ash content and its constituents, for more comprehensive and appropriate assessment of which, it is required to upgrade and improve the available procedures and methods of mineral exploration and sampling. The cost of such modernization seems economically expedient when mining large complex-structure coal beds, panes or deposits.

Based on the evidence obtained with the proposed analysis tools, it is possible to implement coal quality control using different methods (Table 2) [5, 9–13, 16–20]. The list in Table 2 is also incomplete, it should be continuously replenished, ranked and improved. At the same time, even these actions, after the relevant technical-and-economic evaluation and proof, are sufficient to reach more complete extraction of mineral reserves per unit extraction volume.

Table 2. Measures to control coal quality at the stages of mining and preparation of coal in an open-pit mine.

Coal mining and preparation in an open pit mine	$A^d{}_1$	$A^d{}_2$	$A^d{}_3$
Ranking of coal in terms of its preparation to extraction and based on its ash content, considering number and output of production faces	■	■	■
Norming of geotechnical impoverishment	■		
Variation of production face parameters per beds and extraction panels		■	■
Scraping of coal–rock contacts, generation and control of separate coal flows	■	■	
Selective extraction of dirt bands and coal of different quality, and generation of coal flows with more uniform coal quality	■	■	■
Generation of sub-benches, considering coal quality in different layers		■	■
Preliminary cleaning of coal in face areas and intermediate storage yards	■	■	
Change of grain composition by loosening using drilling and blasting		■	■
Employment of surface miners, bulldozers and scrapers	■	■	■
Generation of storage yards to keep coal with the same characteristics	■	■	■
Blending in stationary and "wheeled" storages (various schemes)	■	■	■
Coal grinding in storage yards in an open pit mine	■	■	■
Integration and fusion of coal flows from face areas and from storage yards	■	■	■

Conclusion

The supplementary analyses of the data obtained in additional exploration at Elginsk deposit offered new information on structure of ash content of ill-dressable coal planned for extraction, considering natural and geotechnical impoverishment of coal.

The studies into geometry of coal beds and spatial variability of coal ash content divided into a few constituents have discovered advisability of adjustment of current approaches to coal ash content control within the single geotechnical space generated by Elginsk coal deposit–open pit mine–preparation plant.

The authors have proposed a package of the analysis tools and measures of technological and organizational control over coal flows generated in the course of mining in the framework of coal quality management system to be created and improved at Elginsk deposit.

References

[1] Tkach SM, Batugin SA and Barakaeva ID 2016 Objective assessment of ore dilution—A promising way of enhancing mineral resource potential *Gorny Zh.* No 1

[2] Batugin SA, Tkach SM, Batugina NS and Gavrilov VL 2012 Improvemtn of minineral minin efficiency based on a systemic approach *GIAB* No 9

[3] Batugin SA, Gavrilov VL and Khoyutanov EA 2014 Ash content as a factor to control coal quality in mining of complex-structure deposits *Fund. Prikl. Vopr. Gorn. Nauk* Vol 1 No 1

[4] Batugin SA, Gavrilov VL and Khoyutanov EA 2015 Effects of thin dirt bands on ash content of Elginsk coal *Naukoved.* Vol 7 No 4

[5] Tkach SM (Ed) 2013 *Geotechnologies for Open Pit Mining of Complex-Structure Deposits* Novosibirsk: GEO (in Russian)
[6] Tarzanov IG 2011 Coal industry performance in Russian in January–February 2015 *Ugol* No 3
[7] Kozlov VA 2011 Preparation flow charts for coking coals of Elginsk deposit *GIAB* No 6
[8] Freidina EV, Botvinnik AA an Dvornikova AN 2009 Selecting coal quality and consumer requirements consistency criteria *GIAB* Vol 5 No 9
[9] Ridel RI, Pankevich YuB and Zhakenov BA 2009 New mining technology and equipment in Australia *Gorny Zh.* No 3
[10] Rogova T and Shaklein S 2011 *Reliability of Proved Reserves. Quantitave Assessment and Monitoring* Saarbucken: Lambert Academic Publishing
[11] Freidina EV, Botvinnik AA an Dvornikova AN 2008 Coal quality control in the context of international standards ISO 9000–2000 *J. Min. Sci.* Vol 44 No 6 pp 585–599
[12] Sidorova GP 2006 Methods of dynamic control over Urtui lignite quality *GIAB* No 12
[13] Shchadov VM 2004 *Open Pit Mining of Complex-Structure Coal Deposits in East Siberia and Russia's Far East* Moscow: MGGU (in Russian)
[14] Budge G, Brough J and Knight J 2000 *Process Review of the Worldwide Status of Coal Preparation Technology* RJB Mining Consultancy Ltd
[15] Mohamad Nur Heriawan and Katsuaki K 2008 Uncertainty assessment of coal tonnage by spatial modeling of seam distribution and coal quality *Int. J. of Coal Geology* Vol 76
[16] Benndorf J 2013 Application of efficient methods of conditional simulation for optimizing coal blending strategies in large continuous open pit mining *Int. J. of Coal Geology* Vol 112
[17] Kolesnikov VF Koryakin AI Selyukov AV et al 2010 *Truck-and-Shovel Mining of Rich Coal Zones in Open Pit Mines* Kemerovo: Kuzbassvuzizdat (in Russian)
[18] Shemyakin SYa, Matvee DN and Cheban PYu 2015 Efficiency of cost analysis for no-blast selective mineral extraction and overburden removal using machine groups with cutter–loaders *Gorny Zh.* No 2
[19] Aykul H, Yalcm E, Ediz IG, Dixon-Hardy DW and Akcakoca H 2007 Equipment selection for high selective excavation surface coal mining *Journal of the Southern African Institute of Mining and Metallurgy* Vol 107
[20] Beretta FS, Costa JF and Koppe JC 2010 Reducing coal quality attributes variability using properly designed blending piles helped by geostatistical simulation *Int. J. of Coal Geology* Vol 84

One of the current trends of geotechnology modernization

SM Tkach* **and SA Batugin****

Chersky Institute of Mining of the North, Siberian Branch, Russian Academy of Sciences, Yakutsk, Russia

E-mail: *tkach@igds.ysn.ru; **batuginan@mail.ru

Abstract. The authors describe researches and their results on quality control in mineral wealth reproduction and preservation area. The need and capacity of drastic improvement of geotechnologies and geological support of mineral mining processes given optimization of representative sampling of mineral reserves is demonstrated.

1. Introduction

Natural conditions (physicogeographical and geological) extremely influence efficiency of any branch in the mineral mining industry. They condition the choice of a scheme to get access to a mineral, mining technology and equipment, methods of mineral processing as well as a mine capacity. A consequence of different natural conditions is that technical and economical performance (cost per unit product, labor productivity and other) of mining industry branches may differ by an order of magnitude and more.

The mineral mining depth is one of the key determinants of mining conditions and, thus, technical and economic performance of a mine since the deeper level mining entails elongation of underground openings and transportation networks, increased rate of hazardous events due to rock pressure, complication of ventilation and water drainage, etc. [1–4.

Development of geotechnologies aimed at resource reproduction and saving at all stages of mining is a single, permanent and almost inexhaustible source of enhanced efficiency of the mineral extraction industry.

At the same time, the most complete utilization of that source is impeded by the backwardness of the system of assaying, estimation of mineral quantity and quality and spatial distribution within a deposit [5–8]. This fact is clearly pointed at in the international technical literature, regulatory documents and instruction guidelines.

2. Research findings

Troitsky [5], on the strength of the ample hands-on experience and based on the comprehensive analyses of the specified issue within the latest decades, draws a few conclusions, among which of the interest are:

—actual gold recovery is always higher than the known reserves at placers (Figure 1);

—nobody ever made systematic assessment of the number of sites with proven reserves that were assumed uneconomic based on clearly "procedural" considerations;

—objections to the "uncertainty" of applied procedures are never considered and economic costs of geological exploration with a "negative result" are never repaid and are even neglected.

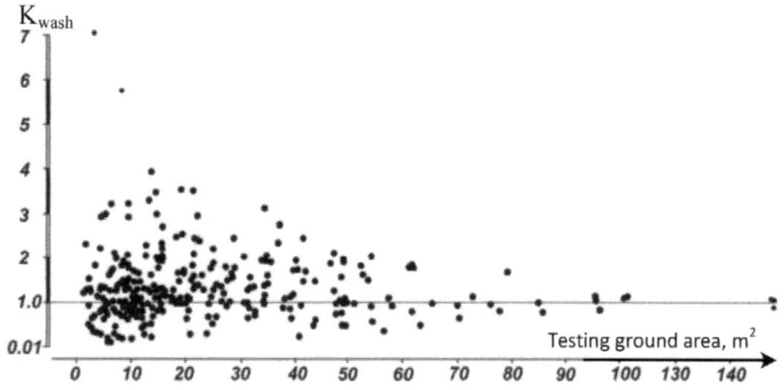

Figure 1. Cheka–Cheka placer mining results.

Jacqui Coombes [6] highlights that:

—two common disbenefits of models of mineral reserves are poor geological justification and insufficient data integrity;

—geostatistics makes a track toward studying and understanding regularities of geological assaying. Nonetheless, without a clear insight into geological evidence, geostatistics turns into the hermetic art in the mathematical wonderland "behind the mirrors;"

—the difficulty lies in combining the qualitative understanding of geological processes and the quantitative data of regularities in order to create reliable spatial prediction that, in the long run, assist mining engineers in efficient mine planning and mineral extraction.

An interesting publication by Ian Ewing in the Russian journal on gold mining [7] presents some recommendations that have been introduced in actual mining:

—special metal disks given a serial number and date are installed at oil stock piles. When a material enters a mill, a magnet elevates the disk, and the data on wherefrom and when the material comes are recorded. Thus, a specific operator who sent the material to a draw hole is identified;

—number of point samples is increased, and lodes are marked with special lines to orient miners and to ensure extraction of ore with the highest iron content;

—the period of sample tests is halved and tonnage of ore with low content is decreased: the next shift miners know where rich ore occurs and where barren rock lies in the beginning of the shift;

—the system of bonuses is changed: bonus is given for the ore product quality rather than for the production and processing output; bonus is given for the production of much gold rather than ore;

—management is learnt to find the golden mean between the cost minimization and labor productivity per miner.

These novations were on the whole discussed with favor and with some doubt about applicability of these recommendations in Russia. However, according to Jacqui Coombes [6], "the Quality Assurance/Quality Control procedures are meant to ensure reliable quality of data used to make decision on relevance of a project and on its progression. The quality of mining and project solutions totally relies on the quality of source data." Unfortunately, in the article and in the discussion, the addressed problem lags behind the state-of-the-art.

Within the late 20 years, the international and Russian science has developed new concepts and new theoretical basis for transition from exploratory constant and variable conditions to operational conditions that are space-differential and time-dynamic. In Russia such transition has been approved by the Ministry of Natural Resources in 1997.

This approach contains considerable resources for enhancing efficiency of mineral mining industry. To understand the scale of these resources, let us discuss operations involved in mineral mining and processing in the inverse order. A feature of processing stage is that ore feed at crushing stage always contains majority of oversizes (value of useful component in a fragment is less than the costs of crushing, milling and concentration to be incurred). For bodies of different minerals and different kinds of the same ore, percentage of oversize in the overall tonnage is different but on the average the mentioned tendency is obvious (see the table).

Percentage of oversized ore in feed of crushing stage.

Mineral deposit	Oversized ore percentage
Uranium	75–85
Tin ore	75–90
Antimony ore	65–85
Iron ore	40–75
Complex ore	30–70
Nonferrous metals	30–70
Rare metals	35–80
Gold ore	40–90
Diamond ore	75–95

Which features of geology and mining technology condition the ratio of quality and oversized ore in an overall output?

First, low thickness of ore lodes (bodies), high variability of thickness and contours, alternation of ore and barren rock layers etc. results in high dilution of ore with barren rocks, which varies in a wide range and most often is 15–50%.

Second, almost for all deposits, a cluster structure is typical, where quality blocks (clusters) make 20–40% of the total ore body and hold 70–90% of useful mineral reserves [8].

Third, the best quality cluster features greatly nonuniform mineralization and highly variable content of useful mineral per small unit area (e.g. 1 m^2), which results in production of much oversized ore (20–30%).

And, finally, fourth, for any ore texture uniformly distributed across an ore body, useful mineral of different dimension is scattered within the body, and any fragments, even of the same size, contain different amount of this mineral. Percentage of oversized ore in this case depends on average mineral composition, grain-size distribution of mineral–bearer of useful component and on the volume of fragments.

Naturally, for any deposit, the structure of mass balance of fragments in broken ore is different. For instance, for diamond deposits, of critical importance are low content and grain-size distribution of diamond crystals, and their rare dissemination in kimberlite. As a consequence, majority of kimberlite from quality clusters is barren. For example, at an African deposit, 50% of samples with a volume of 10 m^3 appear barren. For Yakutia deposits, with much higher content of diamonds, it is characteristics that samples that contain crystals not larger than 0.5 mm make 70% at the sample volume of 0.28 m^3 and 99% at the sample weight of 10 kg.

What is the economical loss of extraction, handling, crushing, milling and concentration of 70–90% of gangue?

It is known that even with the mean distances, ore handling to a processing plant (distance of the order of 15–20 km) costs much more than mining in Yakutia and North–East of Russia. The same holds true for processing. For this reason, an important resource to enhance efficiency of mining industry in these regions is elimination of oversized ore from all operations of extraction, transport and processing.

Implementation of this idea within a mine needs essential changes of basic and auxiliary production processes such as:

—improvement of advanced and concurrent operational and technological assaying in order to detect quality ore clusters within an ore body and to carry out mapping of the ore body and study its morphology;

—development of labeling system for ores of different quality and process properties in production and development stopes with a view to enhancing selectivity of separate extraction of such ores and their subsequent batch sorting;

—introduction of proper and efficient methods and means of geophysical and geological assaying of ore in the host rocks, in bins, flows and fragments and in beneficiation products;

—introduction of new drilling and blasting patterns to be adaptable to selective ore mining and processing;

—change of parameters of production benches (as a rule, reduction in bench height, e.g. to 4 or even to 2 in open pit gold mines) and re-design of underground mines for comfortable selective mining and processing using new process flow charts;

—separation of ore fragments with lower content of useful component at the stages of mining and pre-treatment and sending these fragment for heap leaching or to stock piles to be used later on;

—optimization of the list and parameters of mining machines to conform with the new conditions and requirements of mining, pre-treatment and processing.

The undertaken research has reveled high potential of the listed measures to increase production output per miner by many times at the considerable saving of cost per unit finished product. Conventional notions of loss and dilution are essentially transformed in this case [9].

The basic assaying results are always verified with the more reliable and representative assaying. The comparison of the basic and control assaying data make it possible to substantiate assumability of the basic assaying and to derive some correction factors that are required to be validated as per the regulatory documents.

It almost always happens that the comparison of data obtained in different volume assaying or in assaying using two different methods of different accuracy (representability) yields quantitatively similar conclusions: interval correction factors for the content of minimum size grade more than one (up to 3 or higher) gradually decrease in the higher grade sizes down to figure less than on in the maximum size grades. It was long ago called the "effect of blending" [10].

As regards the method of deriving correction factors: all samples of the basic assaying, properly approved, are grouped into classes of content, and average content $\overline{C}_{0i}$ ($i = 1, 2, \ldots, r$, r—number of classes of content) is calculated per each class. The control sample belongs in the class where the proved sample of the basic assaying is. Based on the proven samples in an i-th class of content, the average contents $\overline{C}_{pi}$. are calculated, too. The correction factor K_{CF} for each class of content is defined by a ratio of the average contents $\overline{C}_{0i}$ and $\overline{C}_{pi}$.: $K_{CF} = \overline{C}_{pi} / \overline{C}_{oi}$.

Let us discuss typical situations that result in various regularities of change in CF with an increase in the number of interval of the basic assaying contents (Figure 2).

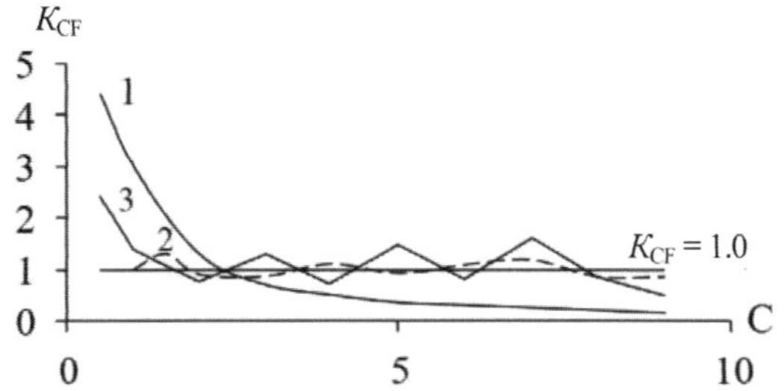

Figure 2. Curves of CF at different combinations of the basic and control assaying: 1—unrepresentative basic and representative control; 2–both basic and control assaying are representative; 3—both basic and control assaying are unrepresentative but the latter is more accurate.

It is known that in a larger sample (or a more representative sample), the interval of probable variation in the content is narrower. Considerably unrepresentative assaying underestimates content more often than not.

The underestimation per unit sample reaches 10 times, which is clearly observed in underground mines and placers of diamond and gold ore as large crystals of diamonds and particle of gold become out of a sample. This feature being typical and known for gold placers and all diamond mines has revealed itself in underground gold ore bodies only in the 21st century [11].

When reserves exploration includes only considerably unrepresentative assaying, out of hundreds and thousands of samples, up to 80% show considerably underestimated content. When we prove these samples using representative control assaying, the result is not underestimated content at lower random error. In accordance with the rule of determination of a correction factor, we put them in the class of the low content together with the proven sample of the basic unrepresentative assaying. As a consequence, CF in the first classes of contents is always much higher than one. It is important that many high values of contents are excluded from the data of the basic assaying.

The last classes of the basic assaying keep a thousandth path of the highest and "hurricane" values. Their average figure is naturally higher than the average of the associated samples, and CF appears less than 1 (Figure 2, curve 1).

Bekker et al [12] give correction factors for vertical margin, average content and thickness of gold sands based on the data of verification of basic assaying in cable drilling in placers in Russia's North–East. The cable drilling was very representative (total 2376 holes). However, as follows from the experience of exploration, appraisal and development of placers of coarse and medium-coarse gold, cable drilling assaying is insufficiently representative.

Percentage of verified holes in the last class of contents as per basic assaying is feeble, 1–3% of the total number of holes. Some of them show actual content in the zone of high contents, and the majority is the abnormally high random values as a result of unrepresentative assaying. In the last intervals of contents, there are 5% of high values of vertical margin in accord with the control assaying.

It seems that at all deposits where the said borehole values have been verified, there are very few holes to show high actual reserves and, consequently, reserves with high vertical margin.

Accordingly, basic assaying shows no sites and areas with high vertical margins, and the correction factor in the classes of high contents (in the unrevealed areas and sites!!!) appears less than one. On the whole, $K_{CF} = 0.98$ per all holes. Nearly half of the basic assaying holes, being considerably unrepresentative show the lowest vertical margin. Here, $K_{CF} = 2.94$.

Correction factor per gold placer was recommended to apply to an entire deposit [13], which is unreflective of a true pattern of reserves distribution per blocks.

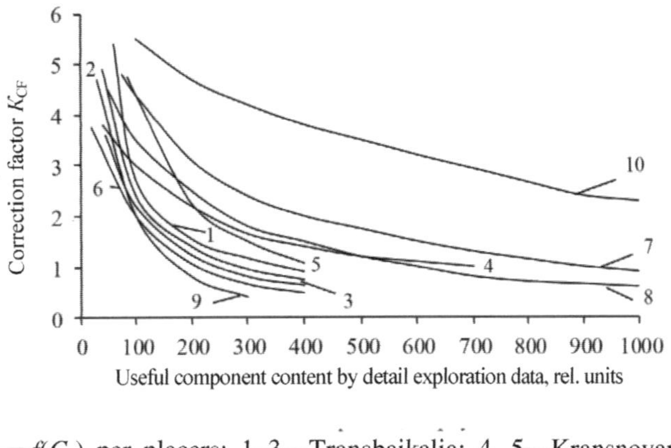

Figure 3. $K_{CF} = f(C_p)$ per placers: 1–3—Transbaikalia; 4, 5—Kransnoyarsk Territory; 6—West Siberia; 7—Lena River valley; 8—Republic of Tyva; 9—Ural; 10—Yakutia. 1–9 are gold placers; 10 is a diamond placer.

Based on the generalization of an immense amount of data on the control assaying of placers, Chemezov [13] offers plots of interval correction factors (Figure 3). CF K_{CF} is determined by the data of bulk assaying of sands per operating dredges, considering assumed loss of the useful component (placers 1–7 and 20) and based on data of the control holes (placers 8 and 9). The plots display the effect of sorting only in the first and third typical situations discussed above (refer to Figure 2) as the basic assaying during the detail exploration of the placers was not representative.

A cardinal difference of an error in the estimate of the useful mineral content of rocks (ores) based on assaying from an error of physical measurements due to geological features of a medium consists in that error of the content estimate can also occur when all direct and indirect measurements in all operation of sampling, processing and analysis are faultless. The error in this case is governed by the rate of representativity of the primary geological assaying. Unallowable random and systemic errors of that kind are possible to avoid only by using representative samples. Erroneous direct and indirect measurements in the course of assaying, preparation and analysis of the samples merely increase the overall mistake.

References

[1] Bovin AA, Kurlenya MV and Shemyakin EI 1983 Problems of extracting deposits of useful minerals at greate depths *J. Min. Sci.* Vol 19 No 3 pp 221–229

[2] Kurlenya MV, Levin VS and Shtelle VI 1983 Theory of mining technology. Prospects of development *J. Min. Sci.*, Vol 19 No 4 pp 317–325

[3] Chersky NV and Batugin SA 1984 Resoruces to enhance deeper level mineral mining efficiency *J. Min. Sci.* Vol 20 No 5

[4] Trubetskoy KN (Ed) 1997 *Mining Sciences. Development and Preservation of the Earth;s Interior* Moscow: Akad. Gorn. Nauk (in Russian)

[5] Troitsky VV 2015 Procedural and legislative issues connected with addition and mining of new gold placer reserves *Zolotodobycha* No 8(201)

[6] Coombes J 2008 *Teh Art and Science of Resource Estmation* Perth: Coonbes Capability

[7] Ewing I http://zolotodb.ru/articles/technical/11143

[8] Tkach SM (Ed) 2013 *Geotechnologies for Open Pit Mining at Complex Structure Mineral Deposits* Novosibirsk: GEO (in Russian)

[9] Tkach SM 2006 *Methodological and Geotechnical Aspects of Improvement in Ore and Placer Mining Effciciency in Yakutia* Yakutsk: Inst. Merzlot. SO RAN

[10] Denisov SA, Arkhipkina TD and Volodin AN 1974 *Issues of Reliabiity of Assaying and Exploration at Ore Deposits* Tashkent: FAN (in Russian)

[11] Konyshev VO, Savostianov EV and Vlasov GN 2003 Fedorovskoe 1 deposit in the Kemerovo Region and Features of Exploration at Deposits Bearing Coarse Gold *Ruda Metally* Nos 5–6

[12] Bekker AG, Tsopanov OKh anb Shevtsov TP 1980 Reliability of cable drilling exploration at placers in the North–East *Geology and Minerals of the North–East* Magadan: Magadan (in Russian)

[13] Chemezov VV Subjectivity in estimation of resources and their utilixation at placers and the impact on the subsoil preservation *Marksheider. Vestn.* No 3

IOP Conf. Series: Earth and Environmental Science **53** (2017) 012017 doi:10.1088/1755-1315/53/1/012017

Effect of georesource–consumer process flows on coal loss in energy supply of the Polar regions in Yakutia

SM Tkach[*] and VL Gavrilov[**]

Chersky Institute of Mining of the North, Siberian Branch, Russian Academy of Sciences, Yakutsk, Russia

E-mail: [*]tkach@igds.ysn.ru, [**]gvlugorsk@mail.ru

Abstract. It is shown that the process flows of mining, haulage and utilization of coal in the Polar regions in Yakutia feature high quantitative and qualitative loss. In case the process flows are considered as integrated systems aimed at the overall performance efficiency, it is possible to reduce the loss per each individual chain loop. The authors formulate approaches intended to lower total loss of coal in process flows. The geotechnical and organizational solutions are put forward to improve and stabilize quality of fuel used by local fuel and energy industry.

The scope of the social and economic development in the regions of Russian Arctic embraces [1, 2]: (1) optimization of the "northern supply" mechanisms, including local power sources; (2) application of resource-saving technologies and up-dated electric power installations; (3) expansion of product range and improvement of quality and competitive ability of products of the local mining industry; (4) maximum utilization of fuel and power resources (FPR) potential. These objectives imply that the current situation in some regions is disadvantageous, and relevant measures are to be undertaken to eliminate the shortcomings. The analyses of all process flows used to supply FPR in adverse and remote Yakutia's Polar regions with the isolated infrastructure prove the statement.

Total 13 Polar municipal regions with an area more than 1.6 Mkm^2 with the population less than 76 thousand residents of 104 settlements consume unreasonably much diesel fuel, benzene, oil, gas condensate, coal and wood per man. At the same time, the federal budget (Yakutia subsidy) and the budget of the Republic of Sakha, which is another name of Yakutia (direct transfer of funds to municipal governments), spend much money to prove the so-called "northern supply." The key producer of heat energy—company ZhKH RS(Ya)—alone consumed FPR in amount of 241.7 thou t, including 51.8% of coal, in 2014. The purchasing (including shipment) cost made Rub 3.77 billion, including 1.40 BRub to purchase coal [3]. The long-term and sometimes critical (up 100% in some areas) dependence on the external supply of FPR, by means of thousand kilometers long, multi-chain delivery schemes, with multiple re-loading stages, under condition of different times of beginning and closure of maritime navigation and winter roads exists so far.

Coal, solely, is partly of "local" origin. The Upper Kolyma district operates Zyryan Open Pit Coal Mine (Kolmar Co.) that supplies coal to population and industry in the basins of the Kolyma and Indigirka Rivers (95 thou t of coal, or 44% of overall coal supply in the Polar regions of Yakutia in 2014). Consumers at the lower reach of the Lena River and in the Yana River basin obtained coal from Dzhebariki-Khaya situated in the Tompon district (119 thou t coal, or 55%). Sakhaenergo company in Ust-Yana district consumed 37 thou t of solid fuel from SUEK's open pit mines beyond the borders of the Republic of Sakha (Yakutia). In 2014, according to the Republic's State Price Policy Committee, the price of coal to be delivered to the adverse northern areas made 5 to 10–12.7 Rub/t (Upper Yana, Ust-Yana, Abyi, Mom districts) and exceeded the world market prices by 2 times and Russian market prices by 3–5 times and above.

This paper addresses the issues of efficiency of process flows between a mineral deposit and a power consumer, where one of the key elements are the large companies listed above, from the viewpoint of loss in the quality and amount of coal in mining, handling and combustion, as well as in the context of assurance of such coal qualities that are required for rational combustion. With a view to ensuring better understanding of the relevancy and urgency of the discussed issues, some argumentation is given below in the paper.

Boiler-house plants in the municipal districts of the Polar zone of Yakutia are conventionally assumed the so-called "small-scale power generation system." These boiler-house plants are the elements of the process flows, where coal produced and delivered from a distance of hundreds and thousands kilometers is converted into heat or, at its best, into electrical energy. These plants use mainly fuel-bed firing and impose higher standards on the quality of a coal product for the fuel-bed firing efficiency, as compared with powdered fuel firing, is to a higher degree dependent on grain-size composition of coal, uniformity of fuel composition, ash content and moisture content and to a lesser degree is dependent on coal grade.

The use of ROM coal with a high content of fine particles instead of standard quality coal product results in the drop of boiler efficiency from 64–73% to 23–35% and in the excess fuel flow by 50–100%. An increase in mine output costs 3–10 times higher than the enhancement of utilization efficiency of classified treated coal [4]. Capital costs of any fuel saving is 4–6 times lower than mining and transport of the fuel taken in the same quantity [5]. Loss due to fine coal combustion or due to its fall through fire bars reach 30% [6]. Moreover, unburned fine coal particles and ash are sent to dumps and continue burning there, which results in toxic emissions of combustion gases in air.

Under combustion of coal from Khabarovsk deposit (Transbaikalia) in boiler-houses and domestic furnaces equipped with 8–15 mm grates, coal particles smaller than 8 fall into ash boxes. On the average, this loss makes 25% in the overall loss of 36%. Also, quality characteristics of this coal (ash content, moisture content, calorific capacity, grain-size composition) have wide ranges, which complicates its combustion in furnaces [7].

By the data obtained at the Skochinsky Institute of Mining, efficiency of a domestic furnace when burning coarse grade anthracite is 64.3%, and when burning anthracite with coal dust, its is 40–31.8%. One of the main causes for dust to appear in a coal product is coal overgrinding. Dust yield makes 35–40% in mine plowing and 45–60% in modern cutting–loading with rotating drum heads. Scraper conveying to distance of 100 m increases the fine size yield by 10% [8], and this is added with 15% by bulldozers commonly employed in open pit mines and coal-storage yards of underground mines [6].

Another problem is connected with the loss of high quality and high energy coal in multi-stage reloading, storage, combustion and distribution of heat and electrical energy. Standard coal loss in transportation within mineral deposit–consumer process flows in the Polar areas is not higher than 8.4–11.3%. However, actual loss, considering the decrease of the energy value, is not less than 10–20%, for instance, in the chain between Dzhebariki-Khaya Mine and Deputat heat power plant, the mentioned loss makes 19.2% [9].

At the same time, sorting plants installed at reloading stations, e.g. Kharanor Open Pit Mine, enable production of standard quality coal, and combustion of such coal improves furnace efficiency to 90% owing to more complete burning. Fine coal particles are recommended for briquetting [7]. The whole-sale prices of coarse standard-quality coal in the Soviet Union were 2–3 times higher than of fine coal, and this remains the same in recent Russia. So, production of coarse coal (at the same output) will essentially enhance profitability of a mine.

The control over generation and yield of coal of various size (together with ash content, calorific effect and moisture content) in the course of mining and primary dressing on-site, with maximum reduction in overgrinding of coal when transported, reloaded and prepared for firing, can improve the performance of the process flows on the whole by means of better and sound utilization of coal potential. Sometimes technical, technological, organizational and economical measures and means require no expensive and high-tech procedures to be implemented, which is of importance for remote areas with underdeveloped infrastructure. The increased yield of fine sizes in underground mining of coal is a complementary argument in favor of open pit mines. Owing to higher efficiency of furnaces when burning standard quality coal, coal shipment volume is reduced and, accordingly, shipment cost is cut-down. The wanted production output of coal can be decreased too.

In this connection, it seems quite logical to introduce notions of theoretical, technical and operating costs of coal, and absolute and relative cost of lost quality of coal in mining and economics [9]. Given an uncertainty of economical and mathematical considerations presented in [9], justification of standard quality coal to be used for fuel-bed firing is acceptable.

Assumable as monopolists, extremely few coal suppliers in the Polar areas of Yakutia pay little attention to coal quality control. This is valid both for mines, which is to a certain degree understandable and explainable, and for the main buyer that is a state represented by authorized consumers of solid fuel, which seems incompletely reasoned.

In Yakutia, coal quality test, with regard to the interests of a producing mine, is executed by the technical services of this mine. This test is a formal procedure, and no receive inspection is performed by a consumer, except for Deputat heat power plant. Quality of supplied coal can "adapt" to changing terms and conditions with time, either because of objective or subjective causes. The former are causes connected with unpredictable aggravation of ground conditions or impoverishment of coal quality in some localized areas (oxidation zones, jointing, thinning of beds, etc). The latter causes are associated with managerial blunder and misdeed, for instance, when "effective" managers validate cutting of expenditures for additional operation exploration. Lack of knowledge on georesource can and results dirt rock cutting in roadway roof and floor due to disagreement of specifications of mining machines and sizes of roadways (e.g. Dzhebariki-Khaya Mine). Overall ash content of produced coal and attle are increased as a consequence. Wide-spaced surveying net and insufficient geological information add to that long-term and current mine planning is performed with no regard to variability of some quality characteristics of coal beds. As a consequence, produced coal has fluctuating quality per different batches.

Mines usually ship ROM coal: 0–300 mm in sizes (Zyryan Open Pit Mine) or 0–200 mm (Dzhebariki-Khaya Mne). Control of quality and the required grain-size composition for boil-house plants is nearly disregarded. Primary dressing (sorting, removal of rock, washing) is absent. Key useful quality (ash content and calorific effect) is low and variable in different batches shipped. Grain-size composition is predominantly represented by fine and very fine sizes, especially in underground mine coal, fuel-bed combustion of which is extremely inefficient. Gross open-pit or underground mining of complex-structure coal beds and impurity of coal with roof and floor dirt rocks lower factual average ash-content of coal down to a minimum allowable level and even beneath it in some product batches. End-to-end monitoring of coal quality variation along the shipment chains is absent.

The causes of that situation are different, including:

—rigid thinking on fuel supply of the remote and power-isolated areas since Soviet period of the state-controlled economy with low, adjustable and state-supported domestic prices of energy sources;

—interest of the parties to preserve the situation as it entails annual, not always rational and transparent budget and off-budget outlays to maintain the "northern supply;"

—possible asking for greater finance in case of various circumstances beyond control, for instance, shallow waters in some rivers;

—traditional approach to coal-driven heat and power engineering, when introduction of internationally and domestically approved, high-efficient and energy-saving technologies, including cogeneration, is obstructed by close-mindedness of a decision-maker;

—absence of real economical incentives and preferences both for consumers and suppliers;

—long-lived money stringency to upgrade all chains in a process flow, for energy-saving activities and improvement of technologies connected with maintenance and enhancement of produced and delivered coal quality.

Complete preservation of current "northern supply" circuits when fuel is delivered to consumers to distances of thousands kilometers, with numerous stages of re-loading, with high cost and loss in the presence of many coal deposits in the Polar regions is inexpedient. This paper authors have reviewed and discussed some scenarios that seem viable, for instance:

—exploration and selection of coal deposits with the best and most attractive characteristics of quality;

—construction of new mini open-pit mines in remote areas of Yakutia, closer to a potential consumer;

—generation of control over coal quality and process flows to ensure efficiency of the entire georesource–consumer system;

—vertical and/or horizontal integration of all mines (process flows) in a single system of efficient coal production and utilization [10];

—introduction of new processes to prepare coal for use, including: mine planning with regard to quality of produced coal, selective cutting; blending, sorting, separate packing of ROM and standard quality coal.

One of the consequences of implementing the put forward scenarios can be new coal supply chains (see the figure) based on mini open-pit mining [3] of the most promising coal deposits [11], for instance, Krasnorechensk field in the Indigirka River basin. As a result, some labor-consuming stages of coal re-loading will be eliminated (e.g. loading from a river vessel to a marine vessel and vice versa); transportation distances will reduce (to 1500 km) and delivery terms will shorten (from 2–2.5 years to a few months); energy preparedness of remote areas in Yakutia will improve.

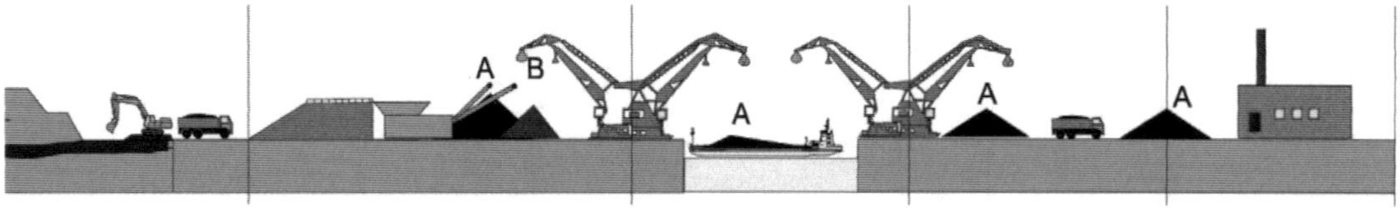

Coal production and delivery flow chain

Introduction of new processes in mines to generate flows of coal of pre-set quality in conformity with the standards of different consumers and to produce standard size ROM coal will greatly decrease considerable total quantitative and qualitative loss of coal, especially in the flow chains connected with logistics and coal combustion. Coarse standard quality coal can be shipped to the most remote consumers in bulk or in bags of 1–2.5 t. Fine size coal is for nearby consumers and/or for dust firing.

Conclusion

The process flow chains in coal supply of the Polar areas in Yakutia feature anomalously high quantitative and qualitative loss of solid fuel, which impairs efficiency of the mining industry and decreases utilization factor of produced coal.

Assuming various process flows as integrated systems aimed to meet a common objective enables reduction in coal loss per individual chains in a process flow. Considerable capabilities in this respect are concentrated in mines—in process flows of mine planning, coal extraction and preparation.

The proposed geotechnical and managerial solutions are aimed to reduce coal loss as well as to improve and stabilize its quality, and are useful in improvement of efficiency of local power generation systems.

References

[1] *Development Strategy for the Arctic Zone of the Russian Federation and National Safety up to 2020* approved by the President of the Russian Federation, Decree No. Pr.-232 dated Feb 2, 2013
[2] *Social and Economic Development of the Arctic Zone of the Russian Federation up to 2020* Government Program approved by the Government of the Russian Federation, Decree No 366-r as of Apr 21, 2014
[3] Tkach SM, Gavrilov VL, Batugina NS, Khoyutanov EA and Fedorov VI 2015 Geotechnical requirements imposed on open pit coal mines in the Polar regions of Yakutia *GIAB* No 7 (Speial Issue)
[4] Zhigulenkova AI, Alekseeva SN and Kekukh TYu 2000 Markets, prices and utilization factor for ROM coals *GIAB* No 2
[5] Nemchinov VP 1983 *Fuel Quality, Efficiency and Price* Moscow: Nedra (in Russian)
[6] Krapchin IP 1976 *Coal Utilization Efficiency* Moscow (in Russian)
[7] Subbotin YuV, Oveshnikov YuM and Tsinoshkin GM 2012 Quality control at Kharanor lignite deposit *GIAB* No 4
[8] Plotnikov VP 2006 Evaluation of hydraulically-driven continuous heading machnies for coarse coal cutting *GIAB* No 3
[9] Zakharov VE, Prokhorov DV and Gavrilov VL 2013 Loss in energy value of ROM coal in shippment to consumers in the Arctic zone of the Republic of Sakha (Yakutia) *Izv vuzov, Problemy Energetiki* Nos 5–6 pp 13–22
[10] Gavrilov VL, Tkach SM, Batugina NS and Fedorov VI 2014 Geotechnical aspects of coal mining in Yakutia's Arctic *Fund. Prikl. Vopr. Gorn. Nauk* Vol 1 No 1
[11] Batugina NS, Gavrilov VL, Khoyutanov EA and Fedorov VI 2014 Coal fields in the Arctic zone of Yakutia and Chukotka: Mineral reserves and their recoverability *Nauka Obraz.* No 4

The problems and prospects of the public–private partnership in the Russian fuel and energy sector

SM Nikitenko[1,3] and EV Goosen[1,2]

[1]Federal Research Center for Coal and Coal Chemistry, Siberian Branch of the Russian Academy of Sciences, Kemerovo, Russia
[2]Kemerovo State University, Kemerovo, Russia
[3]Kemerovo Institute (Branch) of Plekhanov Russian University of Economics, Kemerovo, Russia

E-mail: [1]iuu@icc.kemsc.ru

Abstract. This article highlights some opportunities for shifting the paradigm for the development of natural resources in the Russian fuel and energy sector using public-private partnership instruments. It shows three main directions for developing public-private partnerships in the area of subsoil use and emphasizes the role of innovations in implementing the most promising projects in the fuel and energy sector of Russia.

1. Introduction
The fuel and energy sector (FES) plays a significant role in the development of Russia, with its substantial resources able to meet the country needs and ensure reasonable exports. According to the *BP Statistical Review of World Energy* of June 2016, Russia is ranked second in the world's natural gas reserves, sixth in oil and seventh in coal reserves at the end of 2015. Table 1 shows Russia's share in the world's energy resources in 2015, namely 17% of gas, 6% of oil and 6% of coal [1].

Table 1. Russia's share in the world fossil fuel reserves at end 2015 [1].

Rank in world	Type of fossil fuel	Total proved reserves		Share of total	Reserves-to-production (R/P) ratio
2	Gas (trillion metres cubic***, trillion feet cubic***)	32.3***	1139.6***	17.3%	56.3
6	Oil (thousand million barrels */ thousand million tones**)	102.4*	14.0**	6.0%	25.5
7	Coal (anthracite and bituminous****/ sub-bituminous and lignite*****, million tones	101.5****	14.0*****	6.0%	89.8

Russia is the largest producer and exporter of fossil fuel in the world market. In 2012–2014, Russia ranked first/second in oil production (including gas condensate), sharing the place with Saudi Arabia (12.7–12.9% of world production in 2014), and was the world's second largest oil exporter. It was the second largest natural gas producer (19.6% in 2013, 16.7% in 2014) after the USA (21.4% of world production in 2014) but consistently the largest gas exporter. Russia ranked sixth in coal mining (4.3%) and third in coal exports [2]. According to the Energy Research Institute of the Russian Academy of Sciences, Russia's FES accounted for about 30% of the country's GDP and consolidated budget in 2015. The share of energy resources in total export proceeds amounted to 56% in 2015 [3]. Table 2 indicates that according to the *Forecast of Energy Production in Russia and the World*, the FES will continue to drive the Russian economy up to 2040.

Table 2. Fuel and energy sector in Russian economy [3].

Indicators	2015	2040
Share of fuel and energy sector in gross domestic product (%)	31	15
Share of prime energy in export (%)	56	26
Share of prime energy export in gross domestic product (%)	16	6
Share of fuel and energy sector in consolidated budget (%)	30	14

With its largest resource base, Russia also has a high energy saving potential reaching a third of the current energy consumption, as well as a possibility of significantly increasing the cost-effectiveness of energy projects. However, the exaggerated role of the FES in Russia and its predominantly extensive development has a negative impact on the economy, holding back its development.

The exaggeration of the FES role in Russia has resulted in the formation of a "dual enclave economy," which is most commonly studied in the context of the "resource abundance" and "resource curse"' theories. These terms were used by Sachs, Warner, Gelb and Auty to explain the reasons for the lag in the development of countries with considerable natural resources [4–7]. Later on, Stiglitz linked the problems of "resource-based economies" with the presence of certain enclaves that are quite isolated from each other and use different mechanisms of reproduction [8]. Modern Russian and international studies tend to describe the Russian economy as a "resource-based economy" or a "dual enclave economy" with the features as follows:

- the presence of two or more sectors (enclaves) isolated from each other;
- a high proportion of high-yielding, extensively developing, export-oriented industries involved in the extraction of natural resources;
- a small share of manufacturing industries represented by mainly simple process-based production;
- poor development of industries in the domestic market;
- incomplete processes of market transformation manifested in the underdeveloped domestic market and infrastructure;
- strong dependence on global markets and global, vertically integrated companies;
- low receptiveness to innovative development;
- fusion of business and authorities and their rent-seeking behaviour; and
- the presence of inefficient institutions (rules of the game) preserving the current situation [9-12].

The FES itself is experiencing some negative effects of the enclave economy such as a deteriorated resource base due to the depletion of existing fields, reduced size and quality of new geological discoveries, and increased costs of developing complex and distant provinces. Today, Russia's FES has highly depreciated production assets and outdated technologies, and is too dependent on imported equipment, materials and services, as well as on unstable external energy markets. These problems were exacerbated by a steadily decreasing demand for energy and its prices in the world market, as well as an outbreak of a geopolitical crisis in 2014 leading to the introduction of sanctions against Russia and making it impossible for the country to continue its development by mainly producing natural energy resources and selling them in the external market. There arose an urgent need for seeking a new model of the FES development. The solution seems to lie in a shift from the extensive use of natural resources to an integrated exploitation of natural resources, with sustainable socio-economic development of the regions and mutually beneficial partnership relations between science, authorities and business on the principles of public-private partnership (PPP) [13]. All these factors give rise to a pressing need for seeking potentially productive directions for, and forms of, PPPs aimed at finding a new paradigm for developing the Russian FES.

The purpose of this paper is to identify possible directions for PPPs involving science, authorities and business in the fuel and energy sector that have good prospects in the specific Russian conditions.

2. The problem discussion

PPPs are still quite rare in the Russian FES, although the country's leadership and the *Energy Strategy of Russia for the Period until 2035* emphasise the plans to use PPPs as an effective tool for import substitution and cross-sectoral cooperation that should help to attract investment and "form a domestic scientific, technological and industrial base for designing and producing high-quality power equipment and providing services in the key technological areas to ensure the FES sustainable operation and development" [13]. There are three main reasons for that, namely: (1) Russia does not have a clear conceptual approach to PPPs, (2) the national PPP model is still in its infancy in Russia, with no clear PPP laws, standards and project models, and (3) Russian legislation has a number of serious limitations on using PPP projects in the energy sector.

Among the most authoritative studies of PPPs are those by Osborne, Steven, Klijn, Teisman, Hodge, Greve and Yescombe et al. [14–18]. A few conceptual approaches can be found in the foreign and Russian theoretical studies, differing in their understanding of the PPP role and place in the modern market economy. The first approach takes a broad view of a PPP defining it as any form of cooperation between business and authorities, including joint ventures, corporate social responsibility, charity, even government subsidies, and financial and organizational support for business [19]. This broad approach can show the role of PPPs in the public sector, as well as offer possible directions for joint participation of business and government in the FES development. However, it seems unable to identify PPP features, mechanisms and forms, or their benefits and risks. The main drawback of this approach lies in its inability to clearly define priorities and boundaries for the use of PPPs in the fuel and energy sector.

The second approach is based on the concept of "New Public Management" (van Ham, Koppenjan, Gerrard, Savas, Delmone, Varnavskiy, Deryabina and Kholodnaya) [20–24]. This approach defines PPP as a way of introducing the instruments of commercial project management into the traditional branches of the public sector (medicine, healthcare, protection of public order, public utilities) and of strengthening public property management in the infrastructure sectors, for example, in transport. In other words, PPP is conceptualized as an instrument of state regulation and an alternative to privatization. This approach focuses on adopting foreign organizational schemes and ways of financing projects used in developed countries. Its main advantage is in a detailed analysis of PPPs as a form of interaction between business and authorities, while its main drawback is its focus on the experience of developed countries only. Most followers of this approach do not analyze country-specific PPP models; nor do they consider a possibility of using PPPs in the fuel and energy sector. With some reservations, we can say that this approach is most prevalent in the Russian and foreign literature.

The third approach, most suited for a possibility of using PPPs in the fuel and energy sector, defines PPP projects as an instrument of national, international, regional and municipal economic and social development, and as a way of overcoming the economic crisis in some countries and sectors (Brinkerhoff, Brinkerhoff, Agere, Kontorovich, Nikitenko, Goosen et al.) [25–27]. It clearly defines such PPP features as a project form (a PPP is a long-term project with a clearly defined timeframe), a voluntary and mutually beneficial co-operation between partners, a formal nature of this co-operation based on contracts and agreements with a clear structure of interaction and division of risks and benefits, and a joint participation of business and authorities in the financing, management and/or implementation of the project [12]. At the same time, this approach admits a possibility of PPPs taking the form of a project but not possessing all the PPP features described above. The term "quasi-PPP" ("almost" PPP) has been offered to distinguish such projects from the "classic" PPPs [27]. We believe that it is the third approach that can identify potentially productive areas and specific features of PPP development in the Russian fuel and energy sector.

Let us take a closer look at the Russian experience of PPP development. The Russian market of PPP projects began to formally develop following the Federal Law N 115-FZ "On Concession Agreements" of 21.07.2005. The period of 2005-2015 saw a rapid increase in the number of projects and investment volumes. By the middle of 2016, 1339 projects had been approved for implementation, of which 873 are already underway, according to the Unified Information System of Public-Private Partnerships in the Russian Federation (PPP-info). Table 3 shows a tenfold increase in the number of projects in 2014-2016. It is noteworthy that the Siberian and Far Eastern Federal Districts, where the main resource companies are based, were among those which saw the greatest growth in the number of projects [28].

Table 3. The quantity of PPP projects in Russia as per PPP Information Portal, 2013-2015 [28].

District	Number of PPP projects by year			Growth from 2014 to 2015 (times)
	2013	2014	2015	
Central Federal District	59	21	292	14
North-western Federal District	46	23	114	5
Volga Federal District	36	34	342	10
Southern Federal District	14	10	61	6
North Caucasian Federal District	9	4	28	7
Ural Federal District	25	11	49	4.5
Siberian Federal District	103	24	256	11
Far Eastern Federal District	19	4	143	36
Russian Federation (total)	311	131	1285	9.8

The sectoral analysis shows that most Russian PPPs are based on concession agreements in the infrastructure and social sectors, with very few PPPs set up in the energy sector. In 2015, for example, only five PPP projects out of 1285 belonged to the area of subsoil use, according to the PPP-info database.

All FES-related PPP projects can be divided into three groups. The first group is made up of projects aimed at creating sector-specific mineral productions on the basis of concession agreements and production sharing agreements. Examples of this type of Russian PPPs include development projects in the Elga coal deposit (Republic of Sakha, Neryungri district) and oil and gas fields of Evenkia (Yurubcheno-Tokhomskoye, Kuyumbinskoye, Nizhneangarsk group, and Sobinsko-Teterinskaya group). With some reservations, this group can also include a number of raw hydrocarbons development projects with a share of foreign investment, based on production sharing agreements, namely: Sakhalin-2 Project, including the Piltun–Astokhskoe oil and gas condensate field, and the Lunskoe gas condensate field; the Sakhalin-1 project, including the Chayvo, Odoptu and Arkutun–Dagi oil and gas fields; the Khariyaga project; and the development of the Samotlor oil and gas field. It is noteworthy that such PPPs do not really change the paradigm of subsoil use, retaining the predominantly extensive nature of the FES development and reinforcing the dependency of "host regions" on the extraction of resources and the dual enclave development. However, abandoning such projects at this stage would be unreasonable as they attract foreign investment and can mitigate the effects of the economic crisis, supporting the current state of the FES [31].

The second group of PPP projects focuses on the development of industrial and social infrastructure in the resource extraction regions, creating conditions for a sustainable use of natural resources, deep processing and beneficiation of the extracted minerals.

A good example of such projects is the construction of a complex of refineries and petrochemical plants in Nizhnekamsk initiated by the government of Tatarstan and PAO Tatneft, the main private partner, investor and project coordinator. The project aimed to set up facilities for processing Tatarstan oil near the site of its production; replace oil exports with the realization of high-quality oil products on the domestic and foreign markets, which is in line with Russia's strategic objective; improve the environment by producing environmentally friendly fuels and complying with stringent emission requirements at the design stage; and apply the advanced world technologies. This integration of refineries and petrochemical plants in a single production facility will give impetus to intra- and interregional integration of companies in the region. The first stage of this PPP project has already created over 3,000 new jobs, with new housing, a kindergarten and a sanatorium built for its workers [22, 26]. The distribution of "duties" between the partners is also noteworthy. As a private partner, PAO Tatneft contributed its own funds to the construction of production facilities, while the government of Tatarstan used the Investment Fund (Investfond) to upgrade the external infrastructure such as access tracks, an oil pipeline with a pumping station and a pipeline for the finished products. About 16.5 billion roubles of budget investments was allocated for this project from the Investment Fund, its total cost amounting to over 200 billion roubles. The construction started in 2006 and is now in its final stage [26]. The PPPs of the second type generally have a significant social component. They involve large-scale upgrades in the transport, energy and social infrastructure and aim to improve the social and economic environment by raising employment, living standards, etc. And although they do not produce any fundamental changes in the paradigm of subsoil use, they are much more oriented towards its shift than the first group.

The second PPP group also includes those projects which are components of such mega-projects as Ural Industrial-Ural Polar, the Comprehensive Development of the Lower Angara Area, the Comprehensive Development of South Yakutia, etc. These are long-term multisectoral projects aimed at a large-scale development of new territories. Their implementation involves development institutions and major vertically-integrated companies operating in the energy and mineral resources sectors. All these projects were thoroughly assessed and approved by the RF government. Trans Urals accounted for 62% of total investment in the mega-projects exceeding 150 billion USD, i.e. over 10% of Russia's GDP. They focus on sustainable, balanced and competitive development of problem areas. Table 4 shows some examples of such projects.

Table 4. Russian PPP projects in the fuel and energy sector [29].

No.	Project name	Duration	Industry classification	Project status
1	Elaboration of project documentation for the investment project, "The Comprehensive Development of South Yakutia"	2008-2013	coal mining, enrichment, energy	federal
2	Building the transport infrastructure for the development of mineral resources in the southeast of the Trans-Baikal Territory	2007-2016	enrichment, transportation	federal
3	A complex of refineries and petrochemical plants in Nizhnekamsk	2006-2012	petroleum refining	federal
4	Ural Industrial–Ural Polar	2005-2015	extraction of natural resources, transportation, energy	federal
5	The Comprehensive Development of the Lower Angara Region	2006-2015	transportation, energy, nonferrous metallurgy	regional
6	A complex for processing Northern Caspian gas into ethylene, polyethylene and polypropylene (step I)	2011-2015	gas processing, energy and transport infrastructure	regional
7	Construction of the Kyzyl–Kuragino railway line for the development of mineral resources in the Republic of Tyva	2008-2016	transportation	regional

The Comprehensive Development of South Yakutia project was approved in 2007 under the *Scheme for Comprehensive Development of Production, Transport and Energy in the Republic of Sakha (Yakutia) until 2020*. This PPP project aimed to create a new large industrial zone in the Far East of Russia based on the region hydropower and mineral resources such as natural gas, apatite, coal, iron and uranium ores, etc. It was planned that 25% of its total cost would come from the public funds (Investment Fund) and 75%, from private investors.

The project intended to design and build the following facilities: the Kankunskaya hydropower plant (HPP), the Elkon mining and metallurgical combine (MMC), the South Yakutia mining and metallurgical association, the Inaglinsky coal complex and the Yakutia gas production centre, as well as build roads (Tommot–Elkon MMC, Maly Nimnyr–Kankunskaya HPP), railways (Tommot–Elkon MMC, Taezhnaya–Taezhny mining and beneficiation combine (MBC), Chulbass–Inaglinsky coal complex, Ikabekan–Tarynnakhsky MBC, Kosarevsky–Seligdarsky mining and chemical complex (MCC)), and an electrical grid infrastructure (the Kankunskaya HPP–Neryungri high-voltage line and the Neryungri substation, the Kankunskaya HPP–Aldan high-voltage line and the Aldan substation, and power supplies for the Elkon MMC, the Tarynnakhsky and Taezhny MBCs, the Inaglinsky coal complex and the Seligdarsky MCC).

The project was to remove infrastructure constraints and facilitate the socio-economic development of the region; ensure high rates of economic growth; contribute to an increase in the economically active population in Yakutia, one of the least populated regions in Russia; provide income growth and improved living standards; and increase revenues in the budgets of all levels. It also intended to create conditions for the development of new deep processing facilities, diversify Russian exports, etc. The project had a federal status and, apart from the federal and regional funds, involved such private investors as the Almazredmetzoloto uranium holding, RusHydro, AK ALROSA, EVRAZ, Gazprom, UK Kolmar and other large, national, vertically-integrated companies. A total of 24.8 billion roubles, including over 7.4 billion roubles from the RF Investment Fund, was spent on the Comprehensive Development of South Yakutia, mainly on the design works.

However, despite the enormous investments, the project has not been implemented in full for various reasons, including international economic sanctions, the changes on the world markets of raw materials leading to a sharp drop in the prices for uranium, coal and oil, etc. Yet, the problem seems to lie much deeper. This PPP model—a large investment project involving big companies and development institutions—has been unable to break the inertia of Russia's mineral resource base focused on the extensive use of natural resources. Its experience showed that both businesses and public authorities were most willing to get involved in those parts of the project which were related to the development of new deposits. The private companies built the Inaglinsky coal complex, whereas the government developed its infrastructure.

On the whole, the government contributed 3.2 billion roubles to the Comprehensive Development of South Yakutia invested in the construction of the Chulbass–Inagli railway line, a transmission line and a 110 kV substation. By 2016, the Eastern Siberia–Pacific Ocean (ESPO) oil pipeline had been laid across the south of the republic and the Berkakit–Aldan–Tommot railway had been built. The ongoing projects include the construction of the Power of Siberia gas pipeline and the renovation of Lena Highway. Among the completed facilities are new power lines and a connection to the power grid of the Far East. Much less "lucky" were energy, deep processing and social infrastructure facilities. The issue of expanding the reproduction of the regional resources was also left unaddressed [29].

Even less successful was the Ural Industrial-Ural Polar project launched in 2005. This project was supposed to be able to solve such problems as inadequate investment in the engineering sector, power shortages, underdeveloped energy and transport infrastructure, and the dependence on imported raw materials. A significant part of the investments was assumed to come from the commodity sector through active government policies. The total investment in the project amounted to 543.8 billion roubles in 2006 prices, namely 105 billion roubles from the RF Investment Fund, 79.1 billion roubles from the Ural Federal District, and 359.7 billion roubles from private investors.

The project was supposed to create a unique industrial and infrastructural complex based on the integrated development of natural resources in the Subpolar and Polar Urals, and to construct the key components of the basic transport and energy infrastructure. Its aims were to radically improve the industrial raw materials base, upgrade the transport and energy infrastructure in Russia and the Urals in particular, accelerate the development of regional economies through their large-scale innovative diversification, develop social infrastructure in the new industrial areas, and raise the standard of living and the quality of life for people in the Ural Federal District. The Ural Industrial-Ural Polar project was planned as a model for a new integrated approach to developing mineral resources and territories. It contained three main components—transport, energy, and natural resources. In the core of the transport component was a project to build a Polunochnoye–Obskaya railway line along the eastern slope of the Urals which, together with the Obskaya–Bovanenkovo and Obskaya–Salekhard-Nadym lines and the Salekhard–Agirish–Urai–Tyumen road, was to provide the shortest link between the industrial Urals and the mineral deposits in the Polar Urals and the oil and gas area, ensuring access to the Northern Sea Route to Norilsk. The energy component involved the construction of power lines and four power plants with a total capacity of 2544 MW to run on local raw materials. Its objective was to provide reliable power supplies for the existing and future customers in the new economic zone. The natural resources component envisaged opening 18 to 60 mining, processing and woodworking enterprises. The area under development exceeded 390 thousand sq. km. The project implementation involved over 100 companies and a specially set up management company, OAO Urals Industrial-Urals Polar Corporation, later renamed to AO Development Corporation. The company's shareholders were such regions in the Ural Federal District as Khanty–Mansiysk Autonomous Okrug—Yugra, Yamalo–Nenets Autonomous Okrug, Tyumen Oblast, Chelyabinsk Oblast, Sverdlovsk Oblast, and the German company DB International GMBH. AO Development Corporation prioritized cost-effective investment projects based on PPP principles and capable of increasing the competitiveness of the regional economies, encouraging their diversification and investment activities, removing infrastructure constraints and cultivating new points of economic growth in the Ural Federal District. Its Board of Trustees, consisting of the RF President's plenipotentiary to the Ural Federal District and the governors of Khanty–Mansiysk Autonomous Okrug—Yugra, Yamalo–Nenets Autonomous Okrug and Tyumen Oblast, is responsible for the development strategy of both the corporation and the project [30, 31].

However, just as the mega-project described above, Ural Industrial-Ural Polar has not been implemented in full. Experts believe the reasons for the failure of both mega-projects include their long duration, high risks and complexity, overestimated mineral resources, design flaws in assessing the demand for the transport infrastructure, the economic crisis, sanctions and others. But even if all these negative factors had not come into play and the mega-projects had been fully implemented, they would have been unable to overcome the resource "curse" of Russia and its constituent resource extraction regions [31].

The transition to an integrated, PPP-based subsoil development is only possible if these PPPs aim to set up new production facilities creating innovative development centres, innovative markets and clusters. There are still very few projects like that in the world. According to *IPP Journal*, there are 168 innovative PPP projects in the world in 2016, including 45 in the USA, 10 in Indonesia, and 7 in Bangladesh. Most of the projects are being implemented in the energy sector, constructing modern power plants, with fewer in the oil and coal chemistry. It is noteworthy that private investment for these projects comes from major engineering companies which, together with research organizations and public authorities, establish innovative development centres, technological platforms, etc. [32].

The third group of PPP projects in the Russian FES can be exemplified by three technological platforms involved in the extraction of natural resources and the processing of oil and gas. These three platforms (No. 22, No. 23 and No. 24) were selected from the list of 28 platforms and approved by the Government Commission on High Technology and Innovation in 2011. Platform No. 22 focuses on deep processing of solid minerals and upgrading domestic processing facilities with high technologies. Platform No. 23 aims to develop and implement new technologies for hydrocarbon production, preparation, processing and transportation, including well drilling, etc. Platform No. 24 deals with deep processing of hydrocarbon resources and intends to create conditions for upgrading technologies and increasing the competitiveness of oil and gas processing and petrochemical and organic synthesis, using foresight procedures. The primary goal of these technological platforms is to select fundamental research ideas, do prospecting works and research, and develop innovative business concepts. [26]

The main difference from the first two groups is that the recipients of investments in the third group of PPPs are not producing or processing companies, but those companies which work for the fuel and energy sector, creating internal and external markets, and developing its mineral and human resources, its industrial and scientific base. This limits the possibility for large, vertically integrated energy companies to accumulate all the resources and leads to a synergistic effect, i.e. a large number of highly specialized and competitive small and medium-sized innovative companies setting up around the core PPP project and encouraging the development of scientific and social infrastructure. This ultimately creates conditions for a transition of the FES companies to a rational model of subsoil use, for changing the vector of resource regions development and breaking the vicious circle of the "resource curse."

3. Conclusion

Our analysis shows that the PPP projects in the Russian fuel and energy sector are still developing very slowly. However, PPPs focusing on the innovative model for integrated development of natural resources have potentially great prospects. It is this kind of PPPs that can fundamentally solve the problems of the FES and resource extraction regions through a transition to a new paradigm of subsoil use based on the ideas of integrated use of natural resources and sustainable socio-economic development by adopting new technologies and developing the domestic market.

4. Acknowledgements

This research was supported by the Russian Science Foundation, Project No. 16-18-10182 "Formation of the organizational and economic mechanisms of integrated development of mineral resources based on the partnership between science, business and authorities in the resource-type regions."

References

[1] *BP Statistical Review of World Energy 2016* London Available at: http://www.bp.com/content/dam/bp/pdf/energy-economics/statistical-review-2016/bp-statistical-review-of-world-energy-2016-full-report.pdf.
[2] *State Report on the Status and Use of the Russian Mineral Resources in 2014* 2015 Moscow Available at: http://www.mnr.gov.ru/regulatory/list.php?part=1935.
[3] *Forecast of development of the world energy and Russia in 2016* 2016 AA Makarov et al (Ed) Moscow Available at: https://www.eriras.ru/files/forecast_2016_rus.pdf.
[4] Sachs JD and Warner A M 1995 Natural resource abundance and economic growth NBER *Working Paper* No 5398.
[5] Gelb A et al 1988. *Oil Windfalls: Blessing or Curse?* N.Y.: World Bank Research Publication.
[6] *Resource Abundance and Economic Development* 2001 Auty RM (Ed) Oxford University Press.
[7] Stiglitz JE 2002 *Towards a New Paradigm for Development: Strategies, Policies and Processes* Columbia University: Academic Commons Available at: http:// hdl.handle.net/10022/AC:P:19637.
[8] Ross M 1999 The political economy of the resource curse *World Politics* Vol 51 pp 297–322.
[9] Polterovich VM, Popov VV and Tonis AS 2007 *Economic Policy, Quality of Institutions, and Mechanisms of Resource Curse* Moscow
[10] Levin SN, Sablin KS and Kagan ES 2015 "Resource type" regions in the modern russian economy *J. of Institutional Studies* Vol 7 No 3 pp 92–101.
[11] Rutskiy VN and Pyzhev IS 2015 Institutional problems and development perspectives innovative entrepreneurship in resource economics *J. of Institutional Studies*

Vol 7 No 4 pp. 128–137.

[12] Nikitenko SM, Goosen EV and Sablin KS 2016 Perspectives of comprehensive mineral exploitation based on the principles of public-private partnership *IOP Conf. Series: Earth and Environmental Science* 45 doi:10.1088/1755-1315/45/1/012001.

[13] *Energy Strategy of Russia for the Period 2035* 2016 Moscow.

[14] Delmon J 2009 *Private Sector Investment in Infrastructure: Project Finance, PPP. Projects and Risk* The World Bank and Kluwer Law International.

[15] Osborne S and Steven P 2000 *Managing Public-Private Partnerships for Public Services: An International Perspective* London: Routledge

[16] Klijn EH and Teisman GR 2000 Governing public-private partnerships: Analyzing and managing the processes and institutional characteristics of public–private partnerships *Public-Private Partnerships: Theory and Practice in International Perspective* Osborne SP (Ed) London: Routledge pp. 84–102.

[17] Yescombe ER 2015 *Public-Private Partnership: The Basic Principles of Finance* Moscow: Alpina Publisher

[18] Gerrard MB 2001 Public-private partnerships: what are public-private partnerships, and how do they differ from privatizations? *Finance & Development* Vol 38 No 3 pp 48–51.

[19] Ham H van and Koppenjan J 2001 Building public–private partnerships: assesing and managing risks in port development *Public Management Review* Vol 3 No 4 pp 593–598.

[20] Savas ES 2000 *Privatization and Public Private Partnerships* NY: Chatham House Publishers, Oxford University Press

[21] Delmon D 2010 *Public–Private Partnership in Infrastructure* Astana: Apelsin

[22] Varnavskiy VG 2011 Public–private partnership: some questions of theory and practice *World Economy and International Relations* Vol 9 pp 41–50.

[23] Deryabina M 2008 Public–private partnership: Theory and practice *Vopr. Ekon.* No 8 pp 61–77.

[24] Kholodnaya ND 2009 Public–private partnership—A new type of relations in the Russian economy *Problems of State and Municipal Management* No 2 pp 42–56.

[25] [25] Brinkerhoff DW and Brinkerhoff JM 2004 Partnerships between international donors and non-government development organizations: Opportunities and constrains *International Review of Administrative Sciences* Vol 70 No 2 pp. 253–270.

[26] Agere S 2000 *Promoting Good Governance: Principles, Practices and Perspectives* London: Commonwealth Secretariat

[27] Goosen EV, Nikitenko SM and Pakhomova EO 2015 Experience in the implementation of ppp projects in Russia *EKO* No 2 pp 163–175.

[28] *Unified Information System of PPP in the Russian Federation* Available at: http://www.pppi.ru.

[29] *Reference Materials on the Projects Implemented with the Involvement of the Russian Federation Investment Fund* 2015 Moscow: Ministry of Economic development of the Russian Federation.

[30] Plyaskina NI and Kharitonov VN 2010 *Problems of Formation and Realization of the Mega-Project of the East-Siberian Oil and Gas Complex* Available at: http://isem.irk.ru/symp2010/papers/RUS/S2-04r.pdf.

[31] *The Industrial Ural—Polar Ural Investment Project* 2006 Available at: http://uploadrb.ru/upload/archive/dop_upload/file_2009-12-24_19.44.29_ural.pdf.

[32] *IPP Journal* Available at: http://www.ippjournal.com.

IOP Conf. Series: Earth and Environmental Science **53** (2017) 012019 doi:10.1088/1755-1315/53/1/012019

Technology and means of a coal seam interval hydraulic fracturing for the seam degassing intensification

VI Klishin[1], GY Opruk[1] and AL Tatsienko[2]

[1]Federal Research Center for Coal and Coal Chemistry, Siberian Branch of the Russian Academy of Sciences, Kemerovo, Russia
[2]Kirov Mine, SUEK-Kuzbass, Leninsk-Kuznetsky, Russia

E-mail: [1]klishinvi@icc.kemsc.ru

Abstract. Interval hydraulic fracturing use for the seam degassing intensification actuality is explained. The known methods of degassing are reviewed. Technological scheme of the interval coal seam hydraulic fracturing implementation is worked out. The equipment to fulfill degassing intensification measures is suggested.

Coalbed methane can be recovered from subsoil independently from coal extraction using gas production technology subject to its profitability both as the associated mineral product (methane production) and in degassing process aimed to provide safe labor conditions (methane safety). The latter case makes possible to intensify mining performance, that is, to increase the rate of heading and to increase the coal extraction face output [1–9]. For this reason, there is a paradoxical situation in the coal industry, when the technical abilities of coal extraction at flat seams increase a few times an allowed extraction rate limited due to the gas factor. We must underline at once that we do not consider the industrial methane extraction with, for example, the USA technology, which allows suggesting that degassing is combined with commercial methane production from coal seams [10, 11]. But it is known that in conditions of San Juan basin (this is the basin where 95% of coal seam methane is extracted and their experience is cited by all researchers) gas extraction from the thickest Fruitland coal seams was due to a very high permeability conditioned by the seam cleat nature, so it is practically coal and gas deposit.

From all known time-related degassing methods [3, 5] we can mark out three types: the early one done before the mining work; the preliminary degassing of the worked out seams and the current degassing from the boreholes drilled from underground openings or from the surface. At present at Russian mines they use degassing of a seam under work from preparatory openings (ventilation or conveyer galleries) in every 15–20 m. Still at the seams with low permeability this technology does not give the necessary degassing level. Using the existing degassing methods in Russia they extract from 20 to 30% of the total volume of methane emitted. Due to the small volume of degassing work and the lack of effectiveness of degassing schemes in many gassy mines of Russia the gas barrier to achieving high speeds of the working heading and high output from extraction faces is retained [19].

A further increase in efficiency, technical and ecological safety of underground mining of highly gassy coal deposits are inextricably linked to the development of special measures and artificial methods to stimulate gas recovery from not unloaded coal seams.

The filtration ability of the coal seam is determined by the number, spatial orientation and absolute permeability of its cracks, through which the free methane can move in the filtration mode. To date, the only industry proven way of a coal seam stimulation in order to increase its gas recovery is the seam hydraulic fracturing (GRP) from the surface—the way is not controlled, time-consuming, expensive and it requires a considerable investment of time (up to 3–5 years and more) for a single borehole development. Besides, in order to provide the degassing efficiency of about 50% the period of gas extraction from the seam should be 7.5 years and to achieve efficiency of 60%—10.8 years, and so on.

Given that 90% of methane contained in coal seams is in the adsorbed state, so in order to bring it out of this state it is necessary to impact on the coal seam so that active methane separation would happen. The existing degassing methods when degassing boreholes are drilled from the surface or from underground openings shows that through the borehole walls of the well gas from a very small area of the coal seam in the vicinity of the borehole is released.

Attempts to borrow the known fracturing technologies from oil production face a number of difficulties. Unlike petroleum industry, coal production usually requires the creation of cracks in a direction substantially not coinciding with the direction of the main compressing stress action (for example, inclined plane). To intensify the degassing processes an original way of vibration exposure from the surface of a special vibration sources has been proposed [12, 13], tested on the uranium deposit. Gas releasing ability of not unloaded coal seams can be increased by improving their reservoir properties based on interval hydraulic fracturing method (Figure 1), one of the variants of this idea was presented in [14].

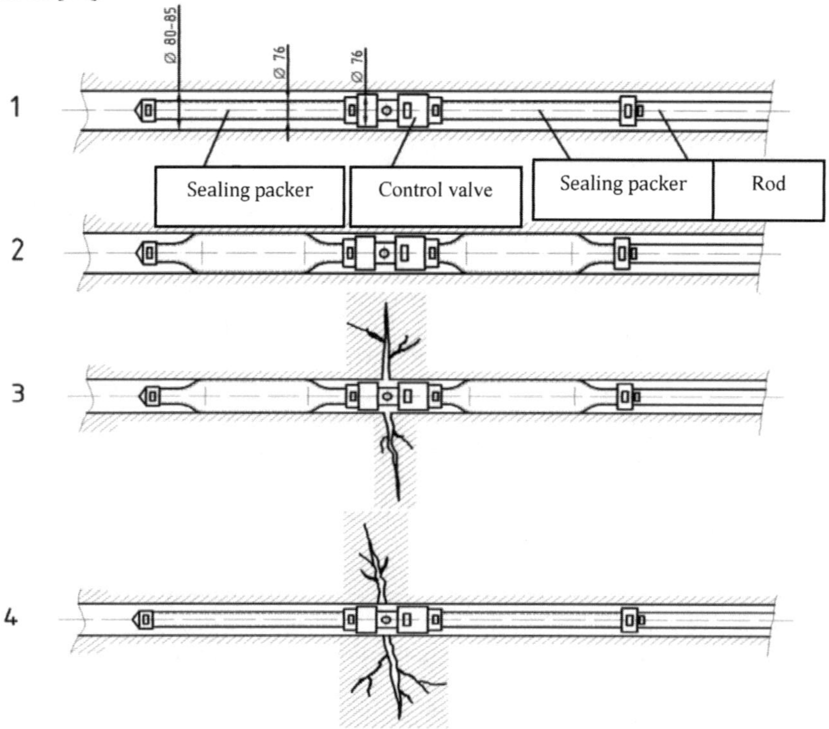

Figure 1. Technological scheme of coal interval hydraulic fracturing: 1—putting the packer into the borehole; 2—sealing of the borehole; 3—hydraulic fracturing of coal; 4—depressurization of the borehole and advancing the packer.

Particular importance is attached to design of the between-the-packers control valve that supplies working fluid for the borehole and two sides sealing and subsequent disruption of the coal seam in the sealed borehole area. Moving packer along the borehole one can make hydraulic fracturing with water or some water solutions at the specified sections of a coal seam. Balanced packer with a new type control valve allows increasing the reliability of sealing on both sides. To implement the coal mass fracturing technology a double sided packer is developed, allowing to carry out the interval hydraulic fracturing (Figure 2).

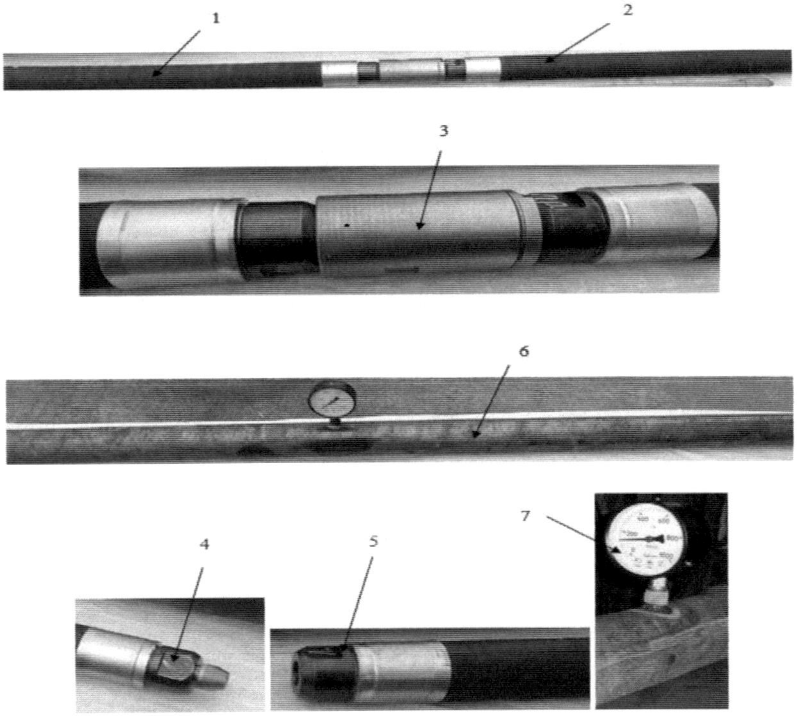

Figure 2. Fracturing device and a coal borehole simulator: 1, 2—elastically expanding sleeve, 3—between-the-packers control valve, 4—locking end cap, 5—connecting coupling; 6—coal borehole simulator; 7—gauge.

The fracturing device includes two elastically expanding sleeves 1 76 mm in diameter, between them a between-the- packers control valve 2 is installed necessary to control the working fluid in the fracturing device hydraulic system. It serves for timely packing the borehole in the zone between the elastically expanding sleeves and for further pressure increase in this area to the value sufficient to fracture the coal massif.

Between-the-packers control valve (Figure 3) consists of a shallow case constructed as a cup 1 with holes 2 for the working fluid passage which is connected to the open part of the hollow hub 3 by means of threads. At the ends of the case elastically expanding sleeves 4 and 5 are fixed which can change their form under working fluid pressure action. Inside the case near the cup bottom a sealing bushing 9 is attached with the spacer 8, wherein the spool 10 is placed and its central opening contains a hollow tube 11 mounted between the bottom of the cup 1 and the end of the hollow hub 3.

In the walls of the sealing bushing 9 there are radial holes 12 axes of which coincide with the axes of the holes 2 in the walls of the cup 1. Pressing spool 10 to the end of the sealing bushing 9 is performed by a spring 13 mounted between the spool 10 and the bottom of the hub 3. At the ends (from the cup bottom side 1) of the spool 10 and hollow tube 11 there are hollow spline slots 14 and 15, respectively. Sealing spacers 16–18 in the hollow hub, at the spool 10 and in the sealing bushing 9, respectively, provide sealing of all the device connections. The surface of the sealing bushing 9 hole is lapped along the outside diameter of the spool 10, which also provides sufficient connection sealing.

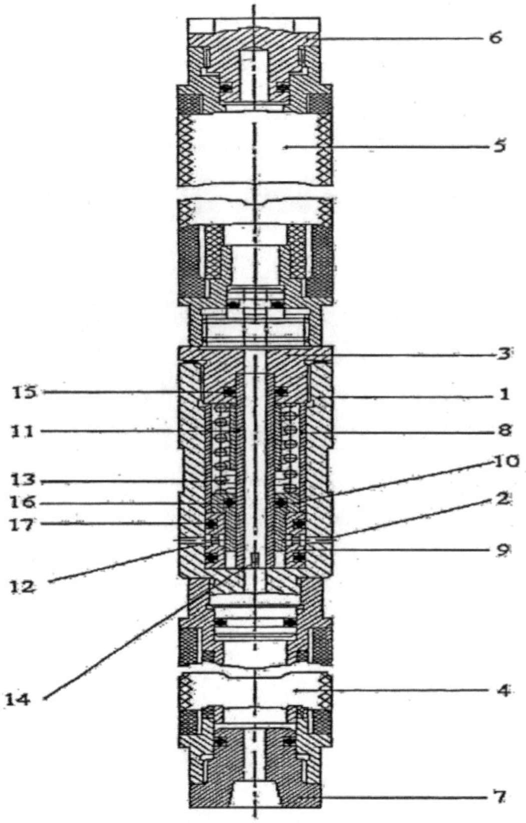

Figure 3. Between-the-packer control valve.

The device works in the following way. Through connection 7 the hydraulic fluid is supplied under pressure to the device, where it acts on the sealing members 4 and 5, increasing in diameter and pressing them against the walls of the borehole—so sealing of its portion occurs. When the pressure to which the spring 13 is adjusted increases, spool 10 starts moving to the hollow hub 3 which provides access to the radial holes 12 and 2 in the walls of the sealing bushing 9 and the cup 1 through which the working fluid enters the sealed space of the borehole. After hydraulic fracturing is done the working fluid pressure is reduced and spool 10 under action of spring 13 returns to the initial position and elastically expanding sleeves 4 and 5 get the initial shape.

Studies of the fracturing device working modes through a between-the-packers control valve were carried out on the test bench (Figure 4).

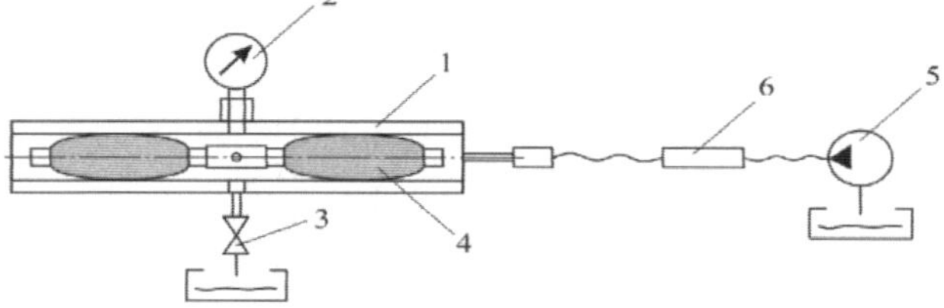

Figure 4. Test bench for testing technologies and tools for a coal seam interval hydraulic fracturing.

The test bench consists of a piece of a pipe 1 imitating the rock massif borehole. The pipe 1 with the inner diameter 80 mm has a gauge 2 and a ball cock 3. Fracturing device 4 inserted into the pipe 1 is connected to the pump station 5 with a pressure pipeline. The pressure pipeline is equipped with Stauf DM-750 flow meter made in Germany with 16 l/min capacity at 32 MPa pressure.

Tests were conducted in the following sequence. The fracturing device 4 was inserted into the pipe 1 in such a way that between-the-packer pressure valve was in the area of the gauge 2 and ball cock 3 installation. Then with the ball cock 3 closed, working fluid under pressure was pumped from the pump station 5 through high pressure pipeline with the flow meter 6 to the fracturing device 4. When the pipeline pressure reached 7 MPa (according to the pump station gauge), pressure in pipe 1 in the area of between-the-packers control valve location also started increasing that was recorded by gauge 2. That meant the fracturing device elastically expanding sleeves isolated the area where the between-the-packer control valve was located and the spool spring 4 (Figure 3) adjusted for 7 MPa compress and allowed the spool 10 working fluid under pressure bypass to the between-the packer area.

The maximum working fluid pressure in the between-the packer area recorded during the test was 16 MPa. There was no fluid leakage around elastically expanding sleeves. The second stage the fracturing device test was conducted to determine the uniformity of the working pressure distributor during the working fluid pumping into the between-the-packer area.

At that time the ball cock 3 (Figure 4), was opening which provided hydraulic fluid free release from between-the-packer area to the surface. Then working fluid was pumped to the fracturing device 4 from pumping station 5 along the pressure line with the flow meter. The amount of working fluid was recorded by the flow meter and the pressure—by a pumping station gauge.

When the pressure in high pressure line was achieving 7 MPa the working fluid was released from between-the-packer area through the open ball cock. That meant the fracturing device elastically expanding sleeves isolated the area where the between-the-packer control valve was located and the spool spring adjusted for 7 MPa compress and allowed the spool to bypass the working fluid to the between-the packer area and further through a ball cock to the surface. The speed of fluid flow through three openings 4 mm diameter in the between-the-packer control valve case was 6.5 l/min. The tests proved that there were no working fluid leaks through elastically expanding sleeves when the fluid was released to the between-the-packer area. In Figure 5 the pressurized hydraulic system pressure change graph is presented depicting the process of working fluid release through between-the-packer control valve.

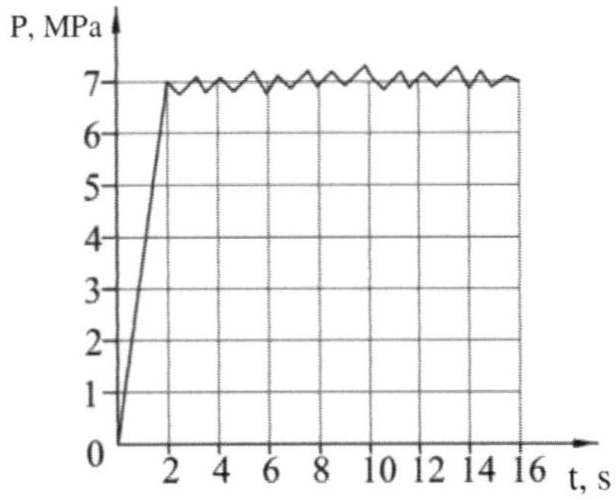

Figure 5. Character of pressure pulsation in the process of working fluid release through the between-the-packer control valve.

The graph shows that pressure fluctuations in the process of the working fluid release do not exceed 0.5 MPa. This slight pressure variation will not affect the reliability of boreholes sealing with elastically expanding sleeves, which are used in a fracturing device.

The proposed technology can be implemented according the following scheme of oriented interval fracturing for gas emission intensification process of methane into degassing boreholes and technical means for its realization (drills, pumping stations, packers) (Figure 6) [15–18]. It allows to avoid the dynamic phenomena in the mines and also to provide safe and efficient work of a coal extraction face with high output. For drilling and interval hydraulic fracturing the drill SBR-400 or SBU-300 is chosen as both have automatic drilling rod replacement without idle running and fast rod returning which increases the work efficiency due to vast time reduction of hydraulic fracturing measures execution.

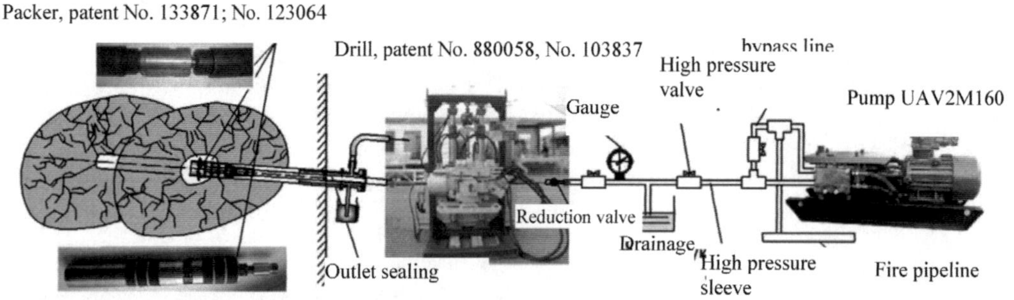

Figure 6. Technological scheme of methane degassing intensification with interval hydraulic fracturing through a degassing borehole.

Conclusion

As a result of the work is expected to create a technology basis for degassing borehole drilling work amount 3 times reduction while increasing the intensity and depth of coal massifs degassing not less than 2 times with the output of methane-air mixtures. Achieving these parameters will allow:

- to increase coal mining safety by increasing the depth of its degassing and reduce the risk of sudden methane outbursts;
- to increase coal mining safety due to outbursts danger reduction with partial unloading of rock massif from stresses using hydraulic fracturing cleats;
- will allow to reduce the amount of degassing boreholes drilling;
- will reduce harmful emissions of methane into the atmosphere.

The laboratory tests proved: fracturing device seals safely between-the–packer area and can be used at interval hydraulic fracturing of coal massif;

Between-the-packer pressure control valve provides timely thrust of elastically expanding sleeves and performs pressurized working fluid supply to the packed area.

References

[1] Trubetskoy KN and Gurianov VV 2003 Underground mining efficiency increase at coal deposits with high gas content on the basis of simultaneous production of coal and methane *Ugol* No 9 pp 3–6

[2] Ruban AD, Zaburdiaev VS, Zaburdiaev GS and Matvienko NG 2006 *Methane in Coal and Ore Mines of Russia: Forecast, Extraction and Use* Moscow: IPKON RAS (in Russian)

[3] Nozhkin NV 1979. Advanced Degassing of Coal Ddeposits Moscow: Nedra (in Russian)

[4] Sergeev IV and Zaburdiaev VS et al 1992 *Gas Emission Control in Coal Mines during Coal Extraction Works* Moscow: Nedra (in Russian)

[5] Slastunov SV 1996 *Advance Degassing and Methane Extraction from Coal Deposits* Moscow: MGGU (in Russian)

[6] Malyshev YuN, Khudin YuL and Vasilchuk MP et al 1997 *Problems of Methane Containing Seams Development in the Kuzbass Coal Basin* Moscow: Akad. Gorn. Nauk (in Russian)

[7] Puchkov LA, Slastunov SV and Fedunets BI 2004 *Prospects of Methane Extraction in The Pechora Coal Basin* Moscow: MGGU (in Russian)

[8] Puchkov LA, Slastunov SV and Prezent GM 2002 Prospects of industrial coal methane extraction]. Gornyi informatsionno-analiticheskii bulleten *GIAB* No 6 pp 6–10

[9] Puchkov LA, Slastunov SV and Kolikov KS 2009 Problems of methane safety concept

realization at Russia's coal mines *Ugol* No 1 pp 28–30
[10] Dugan T and Arnold E 2008 *GAS! History Pages of Coal Methane Extraction in the San Juan Basin* Moscow: CBM Partners Corporation
[11] Sikora P, Smyslov D and Pletner O 2008 Specific features of coal seams advance degassing by drilling boreholes from the surface method *Glukauf* No 1 pp 39–45 (in Russian)
[12] Makaryuk NV, Klishin VI and Kurlenya MV 2002 Physico-technical aspects of rockburst prevention on the basis of vibroseismic impact *Proc. Int. Symp. Rock Bursts-2002* pp 279–288.
[13] Makariuk NV, Klishin VI and Zolotykh SS 2002 Study of mine rocks vibration sensitivity influence on coal seams methane emission at vibration seismic impact *GIAB* No 6 pp 66–70
[14] Klishin VI, Kokoulin DI, Kubanychbek B and Durnin MK 2010 Coal seam weakening as a method of methane emission intensification *Ugol* No 4
[15] Klishin VI, Kokoulin DI, Kubanychbek B Gurtenko AP 2010 Exploration, degassing and service hole drill SBR-400 *Journal of Mining Science* Vol 46 No 4 pp 411–415
[16] Klishin VI and Kurlenya MV 2011 Equipment creation for coal seam degassing on the principle of mine rock fracturing *Ugol* No 10 pp 34–38
[17] Polevshchikov GYa, Kozyreva EN and Shinkevich MV 2012 Efficiency increase of gas emission complex monitoring at a mine coal extraction section *Vestn. Nauch. Tsentra Bezop. Rabot Ugoln. Prom.* No 2 pp 20–27
[18] Plaksin MS, Rodin RI, Riabtsev AA, Alkov VI, Leontieva EV and Nepeina ES 2015 Coal seam hydraulic breakage in mine condition as panacea of the mine gas problems solution (the development and implementation basis *Ugol* No 2 pp 48–50
[19] Nikitenko SM, Goosen EV and Sablin KS 2016 Perspectives of comprehensive mineral exploitation based on the principles of public-private partnership *IOP Conf. Series: Earth and Environmental Science* Vol 45 012001

Structural change and economic growth in modern Russia: The role of "resource-type" regions

SN Levin[1], DV Kislitsyn[1] and KS Sablin[1,2]

[1]Kemerovo State University, Kemerovo, Russia
[2]Federal Research Center for Coal and Coal Chemistry, Siberian Branch of the Russian Academy of Sciences, Kemerovo, Russia

E-mail: [1]levin.sergey.n@gmail.com

Abstract. Authors carry out comparative analysis of economic growth of the subjects of the Russian Federation and highlight their three types: predominance of manufacturing, predominance of services and predominance of mining industries. Based on the results of the research authors make a number of assumptions about the potential of the resource sector and "resource-type" subjects of the Federation as the engines of economic growth.

1. Introduction

From 2011 Russian economy was in the phase of stagnation that showed the inadequacy of the previous model of economic growth based on the export of raw materials and primary processing products of manufacturing industries (metallurgy, chemistry). According to this fact, in the economic literature and projects for economic policy two new challenges have been proposed to the role of the driver of economic growth in the modern Russia. Modern high-tech services are considered as one of the challengers in the frame of the catching-up post-industrialization concept [12, 13]. This approach was mostly reflected in the Strategy–2020. The concept of re-industrialization (new industrialization), which emphasizes the development of the manufacturing industry, is an alternative approach [1, 4].

At the same time one can say that the potential of the traditional Russian resource sector as a potential driver of economic growth is not exhausted. A number of foreign and Russian authors pay special attention to this fact and consider this sector not through the prism of the "resource curse" but as a basis for sustainable economic development [2, 7–9].

The problem of sectoral priorities of economic growth has a strong regional dimension. Reliance on the resource sector puts the focus on the problem of the integrated development of so-called "resource-type" regions, which are the subjects of the Russian Federation that form the backbone of the economy with export-oriented industries and mining industries of the first redistribution. To answer these questions it is necessary to conduct comparative analysis of the regions development with high shares of the manufacturing industry, "service economy" and mining sector. Assessment of the role of the various sectors in the economic growth of the Russian Federation subjects is given according to the data on the development of 80 Russian regions over the period 2005–2014.

2. Theory

The authors of the catch-up post-industrialization concept of the Russian economy consider the fact that within the Russian industrial sector there is no serious advantages, especially if compare it with the newly industrialized countries. Therefore, they believe that the breakthrough strategy should focus on the development, especially high-tech services sector development [14]. It is expected that the accelerated development of high-tech services will create conditions for comprehensive technological modernization of all spheres of the Russian economy. Software development and other computer technologies stand out as the possible competitive advantages of Russia [11]. However, this raises the question about the extent to which the development of such services can become a growth driver of the Russian economy and ensure efficient use of available human, natural and capital resources.

The concept of re-industrialization (new industrialization) is an alternative to the idea of catching-up post-industrialization. At the same time it is based on a long tradition of perception of the manufacturing industry as the "engine of growth" [5]. Empirically, all developed countries had a manufacturing share of employment of at least 18–20% in the second half of XX century. High-growth countries usually experience rapid industrialization. Rodrik [15] showed that manufacturing industry exhibits unconditional convergence that is countries that start farther away from global productivity frontier demonstrates a higher rate of growth, regardless geography, policies, or any other country-specific characteristics.

The engine of growth hypothesis emphasizes capacity of manufacturing to provide special opportunities for capital accumulation and economy of scale. Investments in technology in manufacturing create positive externalities for other sectors, facilitating overall growth. The engine of growth hypothesis closely connected with the dual economy model [10] that implies the existence of two sectors in the economy: more productive modern and less productive traditional. The transition of labor and other resources from former to latter increase average labor productivity in the economy and provide overall economic growth even if there is no productivity growth within sectors.

At the same time the question arises, whether the manufacturing industry keeps its potential engine of economic growth in modern conditions. Those doubts stem from the fact that the de-industrialization has taken large-scale nature in the world economy. This trend is currently affecting not only developed countries but also middle- and low-income economies. Dasgupta and Singh [3] coined the term «premature deindustrialization» to address shrinking of manufacturing sector starting from the significantly lower level of income and manufacturing share that those in developed countries. Rodrik [16] argues that premature deindustrialization reduces the potential for catch-up growth. This raises the question about the extent to which economic policies can provide a way out of the trap of "premature deindustrialization."

After the end of the period of high economic growth in Russia in 2009, public and political debates became increasingly concentrated on the topic of reindustrialization. The anti-crisis package, developed by the Russian government in 2008–2009 included several measures to support the manufacturing sector. For example, The List of Systemically Important Organizations included 111 manufacturing companies, more than any other sector [6]. Largest government loan guarantees were issued for manufacturing firms. Sharp increase in military spending (from 3.4% of GDP in 2008 to 4.2% in 2009, thereafter military spending continued to rise, reaching 4.6% in 2014) was also declared to support the most high-tech part of the Russian manufacturing industry.

At the same time there is a question about the potential of increasing the competitive advantages of domestic manufacturing industry. According to this fact, it is quite reasonable position that the basis for economic growth of the domestic economy may be modernization of its traditional resource industries and creation on their basis value chain covering both manufacturing and services sectors. This approach focuses on the comprehensive development of the resource-type regions. In this paper, we use data on the panel of the Russian regions (formally referred as federal subjects of the Russian Federation) for 2005-2014 years to estimate the impact of structural change on economic growth. In the paper, we employ two measures of sectoral changes: shares of the sector in output and employment.

In the paper, we address to interrelated research questions: (a) did the regions with a larger share of the manufacturing sector in employment and output demonstrate higher rates of economic growth, (b) did the regions expanding the manufacturing sector demonstrated higher rates of economic growth.

2. Data

The primary source of our dataset is Russian Federal State Statistics Service (Rosstat) publications. They contain information about shares of major sectors in output (2004-2014) and employment (2000, 2005-2014) for 80 Russian regions. List of sectors is presented in Table 1.

Table 1. Sectoral shares of output and employment.

1.1. Share of output		1.2. Share of employment	
1.3. Sector	1.4. Abbreviation	1.5. Sector	1.6. Abbreviation
1.7. Agriculture, Hunting, Forestry, and Fishing	1.8. agr	1.9. Agriculture, Hunting, Forestry, and Fishing	1.10. agr
1.11. Mining and Quarrying	1.12. mine	1.13. Mining and Quarrying	1.14. mine
1.15. Manufacturing	1.16. manuf	1.17. Manufacturing	1.18. manuf
1.19. Public Utilities (Electricity, Gas, Water)	1.20. elect	1.21. Public Utilities (Electricity, Gas, Water)	1.22. elect
1.23. Construction	1.24. const	1.25. Construction	1.26. const
1.27. Wholesale and Retail Trade, repair of motor vehicles, personal and household goods	1.28. trade	1.29. Wholesale and Retail Trade, repair of motor vehicles, personal and household goods	1.30. trade
1.31. Hotels and Restaurants	1.32. hosp	1.33. Hotels and Restaurants	1.34. hosp
1.35. Transport, Storage, and Communications	1.36. transp	1.37. Transport, Storage, and Communications	1.38. transp
1.39. Real Estate	1.40. real	1.41. Real Estate	1.42. real
1.43. Education	1.44. edu	1.45. Education	1.46. edu
1.47. Healthcare and social services	1.48. health	1.49. Healthcare and social services	1.50. health
1.51. Other community, social and personal services	1.52. serv	1.53. Other community, social and personal services	1.54. serv
1.55. Finance	1.56. finance	1.57. Other activities	1.58. other
1.59. Government services, military, social security	1.60. state	1.61.	1.62.

The data on shares of the sector in output and employment was complemented with data on the gross regional product (GRP) per capita. We adjusted them to constant 2000 prices using GDP deflators and calculated annual grow rates.

As control variables, we use the index of economic openness of the region, development level of the region, population, and human capital. The index of openness is defined as the share of export plus import in GRP. It was calculated using yearly average dollar exchange rates because data on export and import

IOP Conf. Series: Earth and Environmental Science **53** (2017) 012020 doi:10.1088/1755-1315/53/1/012020

are reported by Rosstat in US dollars. The relative level of development of the region was calculated as the region's GRP per capita as a share of Moscow's GRP per capita (the low value means relatively poor region, variable "rich"). As a proxy for human capital, we used a percentage of employed with university education in total employment of the region (variable "college").

Table 2 presents descriptive statistics of our panel dataset.

Table 2. Descriptive statistics of the panel dataset.

1.63. Variable	1.64. Mean	1.65. Standard deviation			1.66. Observations		
		1.67. Overall	1.68. Within	1.69. Between	1.70. # obs	1.71. # regions	1.72. T-bar
1.73. GRP growth	1.74. 5.45	1.75. 8.65	1.76. 2.02	1.77. 8.42	1.78. 1035	1.79. 80	1.80. 12.94
1.81. rich	1.82. 0.29	1.83. 0.23	1.84. 0.22	1.85. 0.07	1.86. 1115	1.87. 80	1.88. 13.94
1.89. college	1.90. 25.28	1.91. 5.66	1.92. 4.45	1.93. 3.52	1.94. 957	1.95. 80	1.96. 11.97
1.97. openness	1.98. 0.32	1.99. 0.31	1.100. 0.26	1.101. 0.19	1.102. 878	1.103. 80	1.104. 10.98
1.105. ln population	1.106. 7.15	1.107. 0.9	1.108. 0.9	1.109. 0.04	1.110. 1200	1.111. 80	1.112. 15
1.113. Shares of gross regional product:							
1.114. agr	1.115. 9.47	1.116. 5.86	1.117. 5.54	1.118. 2.01	1.119. 879	1.120. 80	1.121. 10.99
1.122. mine	1.123. 7.64	1.124. 12.22	1.125. 11.81	1.126. 3.38	1.127. 879	1.128. 80	1.129. 10.99
1.130. manuf	1.131. 18.3	1.132. 10.62	1.133. 10.25	1.134. 3.02	1.135. 879	1.136. 80	1.137. 10.99
1.138. elect	1.139. 4.78	1.140. 2.58	1.141. 2.39	1.142. 1.02	1.143. 879	1.144. 80	1.145. 10.99
1.146. const	1.147. 7.76	1.148. 3.81	1.149. 2.96	1.150. 2.43	1.151. 879	1.152. 80	1.153. 10.99
1.154. trade	1.155. 14.51	1.156. 5.17	1.157. 4.87	1.158. 1.8	1.159. 879	1.160. 80	1.161. 10.99
1.162. hosp	1.163. 1	1.164. 0.6	1.165. 0.49	1.166. 0.35	1.167. 879	1.168. 80	1.169. 10.99
1.170. transp	1.171. 10.79	1.172. 4.6	1.173. 4.28	1.174. 1.76	1.175. 879	1.176. 80	1.177. 10.99
1.178. finance	1.179. 0.3	1.180. 0.48	1.181. 0.34	1.182. 0.34	1.183. 879	1.184. 80	1.185. 10.99
1.186. real	1.187. 7.29	1.188. 3.21	1.189. 2.87	1.190. 1.47	1.191. 879	1.192. 80	1.193. 10.99
1.194. state	1.195. 7.65	1.196. 4.57	1.197. 4.21	1.198. 1.9	1.199. 879	1.200. 80	1.201. 10.99
1.202. edu	1.203. 4.33	1.204. 1.85	1.205. 1.8	1.206. 0.49	1.207. 879	1.208. 80	1.209. 10.99
1.210. health	1.211. 4.95	1.212. 1.69	1.213. 1.58	1.214. 0.63	1.215. 879	1.216. 80	1.217. 10.99
1.218. serv	1.219. 1.32	1.220. 0.45	1.221. 0.38	1.222. 0.24	1.223. 879	1.224. 80	1.225. 10.99

1.226. Shares of employment:							
1.227. agr	1.228. 12.52	1.229. 5.91	1.230. 5.67	1.231. 1.77	1.232. 877	1.233. 80	1.234. 10.97
1.235. mine	1.236. 1.85	1.237. 2.93	1.238. 2.89	1.239. 0.55	1.240. 877	1.241. 80	1.242. 10.97
1.243. manuf	1.244. 15.16	1.245. 6.2	1.246. 6.11	1.247. 1.33	1.248. 877	1.249. 80	1.250. 10.97
1.251. elect	1.252. 3.33	1.253. 1.49	1.254. 1.46	1.255. 0.33	1.256. 877	1.257. 80	1.258. 10.97
1.259. const	1.260. 7.1	1.261. 2.04	1.262. 1.87	1.263. 0.92	1.264. 877	1.265. 80	1.266. 10.97
1.267. trade	1.268. 15.82	1.269. 3.41	1.270. 3.09	1.271. 1.49	1.272. 877	1.273. 80	1.274. 10.97
1.275. hosp	1.276. 1.65	1.277. 0.52	1.278. 0.46	1.279. 0.25	1.280. 877	1.281. 80	1.282. 10.97
1.283. transp	1.284. 7.91	1.285. 1.87	1.286. 1.8	1.287. 0.54	1.288. 877	1.289. 80	1.290. 10.97
1.291. real	1.292. 6.1	1.293. 2.39	1.294. 2.31	1.295. 0.69	1.296. 877	1.297. 80	1.298. 10.97
1.299. edu	1.300. 9.58	1.301. 2.3	1.302. 2.26	1.303. 0.6	1.304. 877	1.305. 80	1.306. 10.97
1.307. health	1.308. 7.32	1.309. 1.24	1.310. 1.19	1.311. 0.38	1.312. 877	1.313. 80	1.314. 10.97
1.315. serv	1.316. 3.69	1.317. 0.7	1.318. 0.59	1.319. 0.37	1.320. 877	1.321. 80	1.322. 10.97
1.323. other	1.324. 7.96	1.325. 2.28	1.326. 2.18	1.327. 0.76	1.328. 877	1.329. 80	1.330. 10.97

T-bar indicates the average number of observations per region.

3. Methods

We estimate two sets of fixed effects panel regressions: (a) with shares of sectors in output and employment as explanatory variables, (b) with relative changes of sector shares in output and employment. For both sets we separately estimate specifications with shares/relative changes of shares in output and employment. Other variables include log population size, the degree of openness, human capital ("college"), the relative level of development of the region ("rich") and annual time dummies.

As out dataset is a panel, we could take into account unobserved characteristics of the regions by using either fixed or random effects regressions. However, Hausman tests show that region specific effects are not independent of the explanatory variables. The requirement of random effects models is not fulfilled, so we had to opt for fixed effects models.

In our estimation method, we must be careful about interpretation. Statistically significant negative coefficients of relative changes of the sectors shares could not be symptoms of the harmfulness of the sectors for growth, but rather a result of adjustment frictions. Another possible explanation is reverse causality: different sectors respond differently to the recession, some drastically reduce employment (for example, construction), some tend to keep labor (healthcare). Do address this issues we estimate models 1 and 2 years lags of sector variables.

4. Results

Did regions with bigger shares of manufacturing grow faster?

We start by estimating models with shares of sectors in output as explanatory variable (Table 3).

The share of manufacturing is significant with a negative sign in no lag and 1 year lag regressions in both all sample and without outliers specifications. The share of mining is also significant and has an even higher negative coefficient. Services (hospitality, education, healthcare and "community, social and personal services") on the opposite, statistically significant with a positive sign. The important exception is a trade that is statistically significant with a negative sign.

Table 3. Impact of industry shares in total output on GRP per capita growth rate.

	All regions (1)			Without outliers (2)		
Variables	No lag	1 year lag	2 year lag	No lag	1 year lag	2 year lag
mine	-1.145***	-0.740***	-0.794**	-1.156***	-0.671***	-0.876**
	(0.292)	(0.196)	(0.376)	(0.271)	(0.209)	(0.390)
manuf	-0.767***	-0.388**	-0.0613	-0.768***	-0.374**	-0.127
	(0.236)	(0.192)	(0.288)	(0.216)	(0.182)	(0.285)
elect	-0.0134	0.118	-0.0280	-0.0995	0.198	0.0985
	(0.314)	(0.317)	(0.490)	(0.332)	(0.323)	(0.417)
const	-0.396	-0.174	-0.185	-0.382	-0.222	-0.338
	(0.305)	(0.192)	(0.342)	(0.304)	(0.172)	(0.330)
trade	-0.589**	-0.0154	-0.0615	-0.549**	0.0101	-0.116
	(0.277)	(0.175)	(0.356)	(0.271)	(0.186)	(0.347)
hosp	2.068**	-0.995	-0.535	1.782**	-1.177	-0.949
	(0.832)	(0.975)	(1.148)	(0.805)	(0.979)	(1.179)
transp	-0.486	-0.321	-0.206	-0.477	-0.333	-0.273
	(0.296)	(0.315)	(0.471)	(0.294)	(0.288)	(0.480)
finance	0.716	0.740	-0.949	0.245	0.807	-0.584
	(0.998)	(0.715)	(0.728)	(0.925)	(0.706)	(0.691)
real	-0.471	-0.192	-0.320	-0.449	-0.218	-0.466
	(0.370)	(0.260)	(0.445)	(0.380)	(0.256)	(0.451)
state	0.305	0.0187	-0.497	0.373	-0.0113	-0.444
	(0.618)	(0.448)	(0.450)	(0.642)	(0.483)	(0.477)
edu	2.440**	2.348**	0.751	2.794**	2.115**	0.130
	(1.121)	(0.929)	(1.528)	(1.123)	(0.868)	(1.430)
health	1.014*	0.621	0.218	0.930*	0.739	0.536
	(0.578)	(0.643)	(1.001)	(0.537)	(0.665)	(1.013)
serv	2.988*	1.183	1.545	2.505*	0.871	1.324
	(1.551)	(1.770)	(1.511)	(1.495)	(1.754)	(1.573)
openness	-3.562	-0.730	-0.00119	-1.052	-1.037	0.0165
	(2.211)	(4.066)	(4.064)	(3.879)	(4.263)	(4.110)
college	-0.103	-0.0807	-0.153	-0.144	-0.136	-0.198
	(0.108)	(0.124)	(0.176)	(0.118)	(0.138)	(0.176)
rich	68.70***	45.66***	45.50***	58.95***	38.53***	46.82**
	(15.42)	(10.58)	(12.65)	(12.36)	(14.28)	(20.27)
lnpopul	0.867	-0.413	-19.87	-2.044	0.446	-14.87
	(11.74)	(12.07)	(16.08)	(13.24)	(12.92)	(14.91)
Constant	2.595	7.219	155.2	26.09	5.120	127.6
	(86.70)	(84.35)	(123.6)	(96.34)	(91.24)	(117.3)
Observations	719	639	559	692	615	538
R-squared	0.449	0.365	0.296	0.413	0.337	0.289
# regions	80	80	80	77	77	77

*** p<0.01, ** p<0.05, * p<0.1

All specifications include annual time dummies. Robust standard errors are in parenthesis.

 IOP Publishing
doi:10.1088/1755-1315/53/1/012020

Of the control variables only relative level of the region's development is statistically significant, and somewhat counterintuitively it has a positive sign. That means that the gap in economic development among regions increased, as relatively rich regions, grew faster.

Table 4. Impact of industry shares in total employment on GRP per capita growth rate.

	All regions (1)			Without outliers (2)		
Variables	No lag	1 year lag	2 year lag	No lag	1 year lag	2 year lag
mine	-2.285	-3.851***	0.414	0.683	-2.803**	0.461
	(1.585)	(1.298)	(0.604)	(1.483)	(1.200)	(0.748)
manuf	-0.753	-1.089**	-0.603**	-0.631	-0.874**	-0.495**
	(0.545)	(0.451)	(0.260)	(0.516)	(0.416)	(0.245)
elect	-1.091	-1.360	-0.400	1.626	-0.540	-1.808
	(1.394)	(1.582)	(1.503)	(1.692)	(1.499)	(1.554)
const	-0.383	-1.562***	-0.716	-1.248*	-1.832***	-0.795
	(0.866)	(0.568)	(0.603)	(0.717)	(0.658)	(0.574)
trade	0.0003	0.136	-0.881***	0.0454	0.243	-0.883***
	(0.719)	(0.579)	(0.239)	(0.670)	(0.541)	(0.247)
hosp	-1.176	-2.784	-1.551	-1.103	-2.756	-2.168
	(1.920)	(2.160)	(1.879)	(1.880)	(2.225)	(1.858)
transp	-0.582	-2.084***	-0.311	-0.721	-1.869***	0.186
	(0.762)	(0.755)	(0.670)	(0.735)	(0.702)	(0.544)
real	-0.168	-0.684	0.238	-0.210	-1.546**	0.191
	(0.656)	(0.926)	(0.684)	(0.715)	(0.740)	(0.683)
edu	-0.743	-0.221	0.929	-1.101*	-0.439	0.820
	(0.602)	(0.874)	(0.699)	(0.603)	(0.916)	(0.717)
health	-0.378	-1.267*	-0.802	-0.0466	-1.024*	-0.588
	(0.830)	(0.641)	(0.764)	(0.813)	(0.613)	(0.702)
serv	0.469	3.638**	1.736	0.151	2.344*	0.468
	(1.552)	(1.660)	(1.353)	(1.252)	(1.378)	(0.976)
other	-0.712	-0.769	0.768	-0.171	-0.379	0.778
	(0.689)	(0.862)	(0.858)	(0.497)	(0.810)	(0.824)
openness	-2.963	-2.338	-3.252	-3.210	-0.915	-2.220
	(3.996)	(4.475)	(4.017)	(4.425)	(4.399)	(3.691)
college	-0.0879	-0.0792	0.0657	-0.204	-0.225	-0.00918
	(0.133)	(0.171)	(0.192)	(0.137)	(0.183)	(0.165)
rich	23.27	35.19	32.19	12.03	21.63	21.84
	(22.75)	(31.81)	(22.81)	(24.66)	(41.14)	(18.02)
lnpopul	-5.768	8.711	-10.16	5.519	14.19	-6.976
	(15.17)	(16.17)	(13.90)	(15.52)	(18.58)	(14.11)
Constant	90.92	8.361	85.75	4.653	-27.82	72.47
	(121.3)	(112.0)	(100.0)	(120.5)	(126.7)	(102.4)
Observations	638	558	557	614	537	536
R-squared	0.252	0.261	0.193	0.270	0.261	0.201
# regions	80	80	80	77	77	77

*** p<0.01, ** p<0.05, * p<0.1
All specifications include annual time dummies. Robust standard errors are in parenthesis.

Table 5. Impact of relative change in sector shares in total output on GRP per capita growth rate.

	All regions (1)			Without outliers (2)		
Variables	No lag	1 year lag	2 year lag	No lag	1 year lag	2 year lag
agr	-4.681***	-0.518	5.667	-4.337**	-0.463	5.106
	(1.674)	(2.894)	(4.413)	(1.669)	(2.756)	(4.723)
mine	-0.393	0.349	0.439	-0.366	0.348	0.451
	(0.248)	(0.427)	(0.568)	(0.247)	(0.436)	(0.599)
manuf	-3.075***	-1.532	-1.521	-3.204***	-1.607	-0.787
	(1.046)	(1.337)	(1.959)	(1.098)	(1.347)	(1.688)
elect	-4.728**	-2.852	4.477	-4.653**	-3.178	4.993
	(1.806)	(3.199)	(3.165)	(1.790)	(3.200)	(3.203)
const	0.899	0.278	-0.623	0.776	0.203	-0.271
	(1.062)	(2.034)	(1.932)	(1.091)	(2.059)	(1.950)
trade	-3.097	-1.686	-3.179	-2.704	-1.534	-1.717
	(2.069)	(2.465)	(2.970)	(2.063)	(2.496)	(2.839)
hosp	-2.201***	-0.509	-1.631	-2.394***	-0.427	-1.324
	(0.647)	(1.568)	(1.342)	(0.638)	(1.609)	(1.314)
transp	-6.291***	-3.404	1.553	-6.962***	-4.306	3.154
	(2.043)	(3.104)	(3.156)	(2.003)	(3.311)	(3.289)
finance	-0.202	-0.302	0.205	-0.415	-0.429	0.436
	(0.350)	(0.501)	(0.435)	(0.328)	(0.531)	(0.419)
real	-2.940*	-0.471	-0.263	-2.806*	-0.517	-0.355
	(1.564)	(2.355)	(2.746)	(1.540)	(2.422)	(2.760)
state	-16.34***	-1.850	4.888	-16.16***	-0.716	3.900
	(2.814)	(3.796)	(6.935)	(2.953)	(3.920)	(7.221)
edu	-18.53***	3.817	-4.472	-18.22***	2.991	-6.515
	(3.390)	(4.133)	(6.847)	(3.418)	(4.047)	(6.920)
health	-20.03***	6.077	15.27**	-19.89***	6.362	17.70***
	(2.651)	(5.269)	(6.429)	(2.726)	(5.314)	(6.630)
serv	-2.302	-0.793	-3.362	-2.639	-1.152	-3.915*
	(1.807)	(1.943)	(2.031)	(1.847)	(2.029)	(1.985)
openness	-1.905	0.367	-8.655	-2.034	-0.569	-9.607
	(2.758)	(4.818)	(6.163)	(2.899)	(4.856)	(6.183)
college	-0.0320	-0.0590	0.0458	-0.0594	-0.0442	-0.0965
	(0.109)	(0.216)	(0.260)	(0.127)	(0.268)	(0.259)
rich	6.274	7.720	31.11	5.048	4.477	25.14
	(10.32)	(30.47)	(31.39)	(9.095)	(28.84)	(31.17)
lnpopul	15.13	15.70	-0.244	23.33	28.47	8.851
	(18.56)	(26.73)	(29.33)	(18.44)	(26.98)	(35.83)
Constant	-107.0	-105.4	7.786	-166.7	-199.0	-53.89
	(135.1)	(194.3)	(211.3)	(135.3)	(196.4)	(258.2)
Observations	443	370	294	434	362	288
R-squared	0.671	0.277	0.378	0.679	0.290	0.401
# regions	77	77	77	75	75	75

*** p<0.01, ** p<0.05, * p<0.1

All specifications include annual time dummies. Robust standard errors are in parenthesis.

From estimations for the models with shares of sectors in employment as explanatory variable (Table 4), similar conclusions can be drawn for mining and manufacturing (for both specifications 1 year lag statistically significant with negative sign). 1 year lag estimations for construction and transport are also significant with a negative sign. Somewhat surprisingly, estimations for healthcare (1 year lag) and education (1 year lag, only specification without outliers) statistically significant with negative coefficients. That can be explained with inter-sectoral heterogeneity: high-growth regions are relatively rich regions with developed tertiary sector, at the same time, relatively poor low-growth regions are characterized by many low-paid jobs provided by the government in the public sector.

Did regions expanding manufacturing grow faster?

The results for the model with relative changes in sector shares in total output are presented in Table 5.

Estimations for relative changes in nine sectors are statistically significant with a negative sign (agriculture, manufacturing, public utilities, hotels and restaurants, real estate, government services, education). Yet none of 1 lags estimations are significant, so we would be wary interpreting these results as proof that expansion of these sectors reduces potential for growth. One possible explanation is that it is a result of adjustment frictions. If this explanation is true, the more specific capital require the sector, the stronger growth slowdown its expansion causing. That maybe the case with healthcare: it require highly specific physical and human capital. But the expansion of healthcare can increase the growth potential of the region in the long run as evidenced by statistically significant 2 year lag estimate with a big and positive coefficient.

Table 6 shows the results for models with a relative change of sectors in employment as explanatory variables. Estimations for construction and real estate are statistically significant in no lag specification, unsurprisingly as both sectors are pro-cyclical. The only significant lag coefficients are real estate (with a positive coefficient, while it is not significant in the specification without outliers), education and "other activities" (all with negative sign). The negative coefficient of education again emphasizes the same paradox: productive education sector is important for growth, but excessive employment in the sector reduces growth potential.

5. Conclusion

In this paper, we exploited panel dataset on 80 Russian regions for the period 2005-2014 to analyze the relationship between annual shares of sectors in output and employment and per capita gross regional growth. We did not found the evidence of positive impact of manufacturing on growth. Furthermore, our results indicate that regions with a higher share of manufacturing both in terms of output and employment grow slower during the reporting period. The impact of services on growth is paradoxical. We found the evidence that regions with bigger shares of healthcare and education in output growth faster, but a bigger share of this sector in employment, as much as expanding the share of these sectors in employment impede growth. One of the possible reasons for this paradox is that the high proportion of service industries in the regions-leaders (especially in the share of total employment) reflects not the degree of their development but stimulation of the public sector development in the relatively poor subjects of Federation. Composition of the leading regions by the share of sphere of services (Tuva, Chechnya, Republic of Altai, Ingushetia, Kabardino-Balkaria, Chita region, etc.) confirms this fact. The federal government invests substantial financial resources in the public sector, which includes industry data, to encourage development of these regions. As a result, increased share of services reflects the lower level of development of manufacturing and mining industries. Anyway, it is unlikely that the above regions can act as centers of development for the model of catch-up post-industrialization.

Thus, these results show that none of the potential drivers of economic growth has not become a real engine of the regional and the national economy development. This means that the proposed variants of policies to stimulate growth are not used in practice or their tools do not have sufficient efficiency.

Table 6. Impact of relative change in sector shares in total employment on GRP per capita growth rate.

	All regions (1)			Without outliers (2)		
Variables	No lag	1 year lag	2 year lag	No lag	1 year lag	2 year lag
agr	-0.837	-5.831	9.153	4.411	-12.84	-1.709
	(5.334)	(7.101)	(7.533)	(8.157)	(9.625)	(6.849)
mine	0.951	-0.255	-1.533	1.005	-0.377	-1.458
	(1.089)	(1.261)	(1.279)	(1.074)	(1.247)	(1.287)
manuf	15.47	7.130	-0.899	9.640	-5.001	-2.374
	(9.628)	(8.537)	(5.851)	(9.053)	(4.700)	(6.015)
elect	-0.907	1.616	3.753	1.286	2.368	7.624
	(5.868)	(3.505)	(7.188)	(5.877)	(2.785)	(5.772)
const	9.751**	8.338	9.163	9.818**	0.394	-1.793
	(4.559)	(5.623)	(6.679)	(4.667)	(4.930)	(4.708)
trade	10.17	-8.628	-0.772	5.934	-14.21*	-3.432
	(9.500)	(7.458)	(8.739)	(9.959)	(7.970)	(7.519)
hosp	-1.980	-0.541	-0.399	-0.230	-1.366	-2.338
	(2.957)	(3.025)	(2.926)	(3.531)	(2.920)	(3.196)
transp	11.91	7.936	-4.561	6.373	-5.575	-11.32
	(11.01)	(11.45)	(7.360)	(8.208)	(8.973)	(8.174)
real	10.35*	-6.214	7.474*	10.32*	-7.622	2.226
	(5.530)	(5.344)	(4.279)	(5.802)	(6.051)	(3.136)
edu	-40.90**	-33.59**	5.517	-23.72**	-32.05**	-11.18
	(15.87)	(14.86)	(17.22)	(11.39)	(12.46)	(13.08)
health	-10.97	-7.966	-7.701	-13.66	-14.35	-12.22
	(10.14)	(14.19)	(8.619)	(11.22)	(12.84)	(8.007)
serv	-3.004	-11.04	5.520	-2.552	-9.045	9.395**
	(4.809)	(6.902)	(4.364)	(4.925)	(7.197)	(4.089)
other	-8.172	-16.44**	9.099	-6.999	-16.42**	6.789
	(7.794)	(6.889)	(6.930)	(7.809)	(6.897)	(7.002)
openness	-4.537	-4.322	-1.402	-3.193	-2.726	0.226
	(3.263)	(4.400)	(5.542)	(2.911)	(4.242)	(5.252)
college	-0.116	-0.0360	-0.193	-0.190	-0.185	-0.234
	(0.170)	(0.193)	(0.293)	(0.174)	(0.166)	(0.266)
rich	30.20	32.89	70.32**	17.48	24.43	56.79*
	(22.60)	(24.30)	(27.92)	(24.27)	(35.91)	(32.34)
lnpopul	3.488	8.322	-22.39	10.81	20.62	-18.28
	(18.01)	(24.40)	(20.94)	(16.83)	(24.26)	(20.77)
Constant	-19.06	-53.68	149.8	-66.82	-136.0	126.4
	(124.9)	(169.1)	(148.3)	(118.1)	(169.6)	(146.0)
Observations	599	525	450	578	507	435
R-squared	0.288	0.242	0.181	0.298	0.249	0.167
# regions	79	79	79	76	76	76

*** p<0.01, ** p<0.05, * p<0.1

All specifications include annual time dummies. Robust standard errors are in parenthesis.

According to this fact, it appears that the question of the potential development of "resource-type" regions remains open. However, its implementation requires the use of both federal and sub-national policy instruments of smart specialization focused on the search for new competitive advantages associated with the formation of value chains based on the resource sectors of the economy.

6. Acknowledgements

The research was supported by the Russian Science Foundation, Project No. 16-18-10182 "Formation of the organizational-economic mechanisms of integrated development of mineral resources based on the partnership between science, business and authorities in the resource type regions."

References

[1] Amosov A 2009 *Ekonomist* No 6 pp 14–29
[2] Brich A 2003 *Voprosy Ekonomii* No 5 pp 19–41
[3] Dasgupta S and Singh A 2006 *Manufacturing, Services and Premature Deindustrialization in Developing Countries: A Kaldorian Analysis* Available at: https://ideas.repec.org/
[4] Inozemtsev V 2010 *Ekonomist* No 11 pp 3–15
[5] Kaldor N 1966 *Causes of the Slow Rate of Economic Growth of the United Kingdom: an Inaugural Lecture* Cambridge: Cambridge University Press
[6] Kislitsyn D 2013 *Voprosy Ekonomii* No 6 pp 84–99
[7] Krukov VA 2013 *EKO* No 8 pp 4–5
[8] Krukov VA, Tokarev AN and Shmat VV 2014 *EKO* No 3 pp 5–30
[9] Krukov VA 2015 *EKO* No 2 pp 62–76
[10] Lewis WA 1954 *The Manchester School* Vol 22 No 2 pp 139–191
[11] Lynch AC 2002 *Europe-Asia Studies* Vol 54 No 1 pp 31–49
[12] Mau V 2006 *Ekonomic. Polt.* No 2 pp 30–46
[13] Mau V 2007 *Voprosy Ekonomii* No 2 pp 4–26
[14] Mau V 2008 *Voprosy Ekonomii* No 2 pp 4–26
[15] Rodrik D 2013 *The Quarterly Journal of Economics* Vol 128 No 1 pp 165–204
[16] Rodrik D 2015 *Premature Deindustrialization* At: http://www.nber.org/papers/w20935.

Development of chemical compositions for impervious screens in rocks

MV Kurlenya, SV Serdyukov, TV Shilova and AV Patutin*

Chinakal Institute of Mining, Siberian Branch, Russian Academy of Sciences, Novosibirsk, Russia

E-mail: *andrey.patutin@gmail.com

Abstract. The paper presents the method to create anti-seepage screens by hydraulic fracturing with three-component polyurethane mixture. The proposed working fluids and their pumping circuits allow creation of a fracture and an adjacent insulation layer. Gas permeability of porous medium is determined at limit consumption of reagents per insulating screen unit area.

Impervious screen employment as protection of underground working increases mining safety. Impervious screens can be used to seal embedded degassing boreholes [1, 2], to prevent accidental steam entry in mine oil production [3] and underground leaching of non-ferrous metals [4, 5]. Formed in rock mass, the barrier contains a hydraulic fracture and adjoining area with the artificially decreased permeability. It is possible under rock masses treatment with insulating composition which has low initial viscosity and easily permeates into existing interstices and cracks [6].

One-component polymeric compositions and sealants cannot solve most of the mining engineering problems due to high viscosity, weak ability to penetrate and high hydraulic resistance during injection. At the same time two-component fluids possess the necessary properties. They are based on polyurethane resins and acrylate jells with independent supply of low-viscosity components into an impervious screen by separate channels. The mixture of components and edge water causes chemical formation with flexible polymer [7].

Compositions existing on the market used for insulating works are usually based on two-component acrylate jells. The beneficial property of the jells is low viscosity commensurable to water viscosity and lack of organic solvent.

The polyurethane resin components ability to penetrate is weaker than the ability of acrylate jells. However, the weaker ability to penetrate is an advantage for the formation of impervious screens because it decreases leakage to rock masses without injection problems. Due to cheaper reagents polyurethane compositions are more economically beneficial than the acrylate ones.

Most of the polyurethane compositions used for water shutoff have a form of high-viscosity (up to 5–6 Pa·s) prepolymers. They are obtained as a result of isocyanate and polyol component mixture before the work start. Catalyst composition is fixed and baking time of resins is not regulated. These compositions are not suitable for the formation of impervious screens due to viscous prepolymer is insufficiently capable to penetrate and completely fill hydraulic fracture.

There have been developed an insulating composition containing three working fluids, one of which is hydraulic fracturing fluid, two others are polyurethane resin components. The components are low-viscosity prepolymer with isocyanate redundancy (further, prepolymer) and hydroxyl plasticizer with catalyst (further, plasticizer) [8].

During the works performed in an isolated area of rock masses near a hydraulic fracture, it is necessary to follow the described order of injection:

1) At the first step the working fluid is supplied into fracture interval until a crack of a required size is formed in rock. Then, it takes some time till the pressure in the closed fracture interval declines to the formation pressure. The crack surfaces are closed, and the injected fluid permeates the rock mass;

2) At the second step a part of plasticizer (20–30% from the planned volume of prepolymer) is injected into the hydraulic fracturing crack, and it forms a layer saturated with this composition;

3) At the third step the prepolymer and the rest of the plasticizer (volume ratio 1:1) are supplied into the crack by two separate high-pressure hoses. They mix in the fracture interval and penetrate in enclosing rocks through the layer previously saturated with plasticizer. Contained in the layer, diol is reactive with free isocyanate, which leads to the increase of the prepolymer viscosity.

To estimate capabilities of the developed insulating composition and its influence on porous medium permeability, laboratory tests have been conducted.

The following compositions of samples have been used to model the porous medium: abrasive material—white corundum (classification group 24A), medium-soft (classification group SM1 by GOST 19202–80), medium structure (classification group 7 by volume content of the abrasive material), 25–315 micron grain (classification group 25 by GOST 3647–80) with ceramic bond (classification group *K* by GOST 2424–83). From the abrasive wheel, we have drilled and cut cylindrical samples 30 mm in diameter and height. The porous medium model is made of two samples end-conjoined and placed into rubber cover. The insulating composition fluids have been put on the contact surface between the samples with specific charge 0.14–0.84 ml/cm^2.

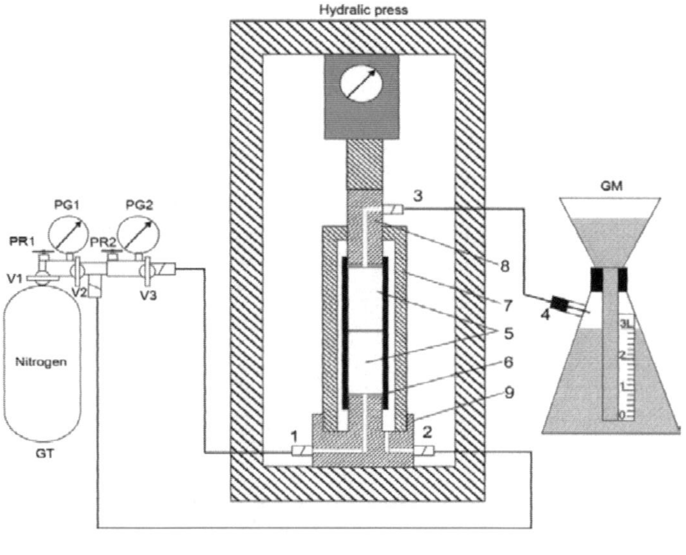

Figure 1. Scheme of bench to study insulating composition influence on porous medium permeability: GT—compressed gas tank; PG1 and PG2—pressure gauges; V1–V3—valves; PR1 and PR2—pressure reducers; GM—gasometer; *1–4*—inputs and outputs for the connecting hoses to plumb in; *5*—porous samples; *6*—rubber cover; *7*—test cell body; *8*—loading rod; *9*—basement.

The study of insulating composition influence on porous medium permeability has been carried out in accordance with the following technique:

1) Setting the prepared model of porous medium in the test cell;

2) Compression the model along its axis on the press with power 2 kN;

3) Opening of the valves V1–2 (Figure 1) and reducer PR1, using pressure gauge PG1 2.5 MPa, lateral compression of the model;

4) The required nitrogen drop of pressure ΔP between the test cell outputs 1 and 3 is set by reducer PR2 using pressure gauge PG2;

5) Water in amount of 3 dm^3 is poured into gasometer GM through funnel;

6) Opening of the valve V3 and observation of water being displaced with gas from gas meter into the funnel;

7) When water reaches "2.5 l" gasometer GM scale mark timer is turned on;

8) When water reaches "1.0 l" scale mark turn off the timer, close the valve V3 and stop the experiment;

9) According to the timer readings filtration time is determined, and in accordance with thermometer and barometer temperature *T* and air pressure in the laboratory are determined.

Figure 2 shows schemes of porous medium treatment with different working fluids employed in the experiments. We have studied the models saturated with water (Figure 2a), hydraulic oil HLP–68 with 175 mPa·s viscosity in 24.8°C (Figure 2b), composition of prepolymer and plasticizer with 2.4 Pa·s (Figure 2c) and insulating composition with catalyst concentration in plasticizer is equal to 0.6% (Figure 2d).

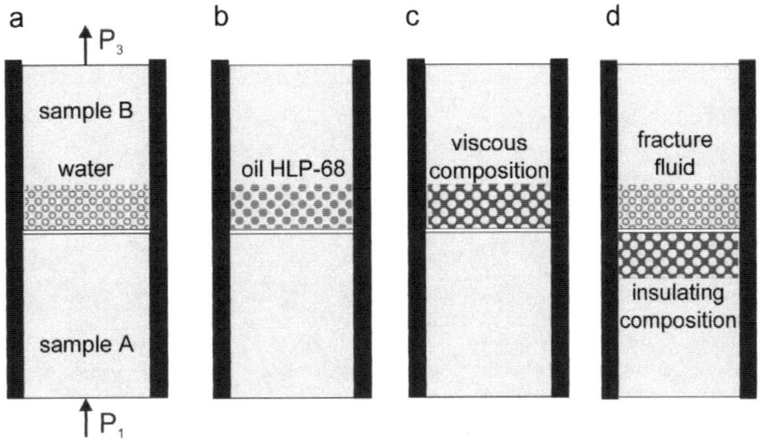

Figure 2. Scheme of studied sample treatment with the following compositions: a—water, b—oil HLP-68, c—composition of prepolymer and plasticizer, d—fracture fluid with insulating composition.

The value of gas permeability coefficient K_g is calculated from the formula for gas linear flow and static filtration [9]

$$K_g = \frac{2 \cdot 10^4 \cdot V \cdot \mu_a \cdot P_3}{t_{av} \cdot S} \cdot L \cdot \frac{1}{\left(P_1^2 - P_3^2\right)},$$

where K_g—gas permeability coefficient, 10^{-3} mcm^2 (mD); $P_1 = P_3 + \Delta P$—pressure at the test cell input *1* (Figure 1), 10^{-1} MPa (bar); P_3—pressure at the test cell output *3* (atmospheric), 10^{-1} MPa (bar); *V*—gas volume with pressure P_3, gone through the model cm^3; μ_a—nitrogen viscosity, mPa·s (sP); *S*—sectional area of the model, cm^2; *L*—model length, cm; t_{av}—filtration time, averaged by several dimensions, s.

The experiments are carried out at $T = 24.8°C$ and nitrogen viscosity of 0.01778 mPa·s [9].

In the course of the experiments it is found that insulating properties of the composition with 0.6% catalyst concentration in plasticizer stabilize in 20–25 min after polymerization is started. It conforms with the cure time of the same composition under room temperature.

Figure 3a shows a graph of porous medium model gas permeability coefficient (treatment scheme in Figure 2d) depending on nitrogen pressure drop ΔP after the insulating composition cure in porous sample. The time after the composition is made is $t_{pm} \geq 20$ min. Figure 3b shows K_g as function of drop of averaged gas pressure in the model $P_{av} = P_3 + \Delta P/2$ implemented for extrapolation of measurement results on gas permeability in range of small values ΔP [10].

According to the graph in Figure 3a gas permeability K_g has non-linear dependence on the pressure drop ΔP. When the pressure drop is low gas permeability increases. In actual practice of degassing borehole shielding, gradient of pressure between the mine working and methane extraction zone is about 100 Pa/cm, which corresponds to $6 \cdot 10^{-4}$ MPa pressure drop for 6c m-long model.

The linear trend of $K_g(1/P_{av})$ dependence shown in Figure 3b is described by the expression $K_g = a/P_{av} - b$, where $a = 362.963$ mD·bar, $b = 89.934$ mD. Determination coefficient is $r^2 = 0.98$. Using the formulated expression we obtain the following value $K_g = 237$ mD (10^{-3} micron^2) for $\Delta P = 0.25 \cdot 10^{-1}$ MPa (bar) and $t_{pm} \geq 20$ min. The obtained value is significantly lower than the measured one in case of incomplete cure of insulating composition (t_{pm}=5–15 min. For ΔP=$0.006 \cdot 10^{-1}$ MPa we have $K_g = 278$ mD (10^{-3} micron^2).

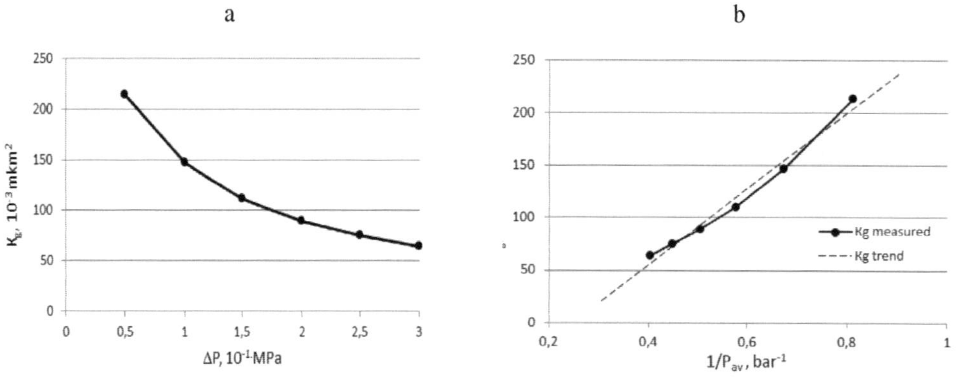

Figure 3. Dependence of gas permeability coefficient K_g on: (a) nitrogen pressure drop $\Delta P = P_1 - P_3$; (b) reciprocal to averaged nitrogen pressure in model when $t_{pm} \geq 20$ min.

Comparing gas permeability coefficient for different treatment schemes (Figures 2a–2d) when $\Delta P = 0.25\ 10^{-1}$ MPa and $\Delta P = 0.5\ 10^{-1}$ MPa we have determined that insulating composition with the specific charge of 0.28 ml/cm^2 (without considering fracture fluid based on water) decreases porous medium permeability more than water by 7–9 times. It decreases the permeability of liquid with viscosity higher than 175 mPa·s and the same charge by 6–8 times, liquid of the same viscosity with 3 times bigger charge—by 5–6 times, and liquid with viscosity of 2.4 Pa·s—by 4 times.

To conclude with, the main role of insulating composition in an impervious screen is to decrease leakage of viscous fill fluid from the crack into enclosing rocks. The developed composition provides the decrease in porous medium permeability as well as in fill fluid leakage by several times in case of low charge per unit screen area.

Acknowledgements
The research was carried out in Institute of Mining SB RAS and supported by the grant of the Russian Scientific Foundation, project № 15-17-00008

References

[1] Polevshchikov GYa et al 1998 Pat 2108464 RU Method for sealing of degassing well *Buill. Izobret.* No 31

[2] Kurlenya MV, Shilova TV, Serdyukov SV and Patutin AV 2014 Sealing of coal bed methane drainage holes by barrier screening method *J. Min. Sci.* Vol 50 No 4 pp 814–818

[3] Shilova TV 2016 Underground workings protection from steam breakthrough during mine heavy oil production *Proc. Problems of Geology and Subsurface Development* Vol 2 pp 868–870.

[4] Vorobiev A,E Chekushina TV, Pogodin ML, Odincova ES 2000 In situ leaching of gold in Russia *GIAB* No 12 pp 45–48

[5] Kurlenya MV et al 1988 Pat. 1368429 RU Underground leaching of rock mass *Byull. Izobret.* No 3

[6] Kurlenya MV, Serdyukov SV, Shilova TV, Patutin AV 2014 Procedure and equipment for sealing coal bed methane drainage holes by barrier shielding *J. Min. Sci.* Vol 50 No 5.

[7] Saunders JH and Frisch CC 1968 *Polyurethane Chemistry*

[8] Serdyukov SV, Shilova TV and Drobchik AN 2016 Polymeric Insulating Compositions for Impervious Screening of Rock Masses *J. Min. Sci.* Vol 52 No 4

[9] GOST 26450.2-85 1985 *Rocks. Method for Determination of Absolute Gas Permeability Coefficient by Stationary and Non-Stationary Filtration*

[10] Peskov AV and Olchovskaya VA 2010 Determination of rock gas permeability with account of gas sliding effect *Oilfield Engineering* No 3 pp 10–12

Application prospects for impact machines in the area of strengthening of joint welds

AD Abramov and MS Galay*

Siberian State Transport University, Novosibirsk, Russia

E-mail: *galayms@stu.ru

Abstract. Effect of vibropercussion plastic strengthening on microhardness and depth of strengthening of tread areas of railway welds is experimentally studied. It is found that improvement of hardness of railway welds requires impact treatment of welds with a unit blow energy of 20 J by an indenter with a cylindrical flat end 55 mm in radius. Under the specified energy impact, a surface layer is formed with the microhardness of 5800–6100 MPa and the strengthened thickness of 1.8–2 mm. The implementation of the technology for vibropercussion plastic strengthening of railway welds needs a moble unit with an electromagnetic drive to be designed.

Machine engineering strives to improve quality of machine parts. An interesting way of enhancing performance of machine parts is hardening when on manufacture and reconditioning. Special operations of hardening include thermal, mechanical, chemical or combination treatment of parts.

For reasoned selection of hardening techniques, it is required to compare capabilities of different methods, in particular, in terms of microhardness and stress state of the hardened layer. On the evidence of the compared capabilities of different hardening methods, the surface hardening of thermostrengthening steels allows microhardness up to 7500 MPa in the strengthened layer not more than 5 mm deep. Stresses in the hardened layer, depending on the chosen thermal treatment mode, can be both tensile and compressive. Noteworthy is the jump from the hardened layer to the nonstrengthened core. After chemical processing, microhardness can reach 8000 MPa with the strengthened layer from 0.01 to 1.00–1.40 mm thick. The sign of residual stresses and their behavior after chemical–thermal treatment depend on the subsequent heat treat process. Hardsurfacing overlay is intended to enhance corrosion stability and wear resistance under friction and abrasion of the surface of parts. Hardsurfacing overlying allows the surface microhardness up to 3000 MPa regardless initial hardness of the part under treatment. The hardsurface overlay thickness ranges from 0.003 to 2.0 mm. Application of hardsurfacing overlay induces tensile stresses in the subsurface layer of parts. Surface plastic deformation SPD enables strengthening of a surface layer to 30 mm thick and increases microhardness on the average by 150% as compared with the initial value. Residual stresses after SPD are always negative and propagate to a depth 1.5 times exceeding the depth of the increased microhardness [1–4].

In this respect, the method of surface plastic hardening is an efficient approach to improving performance of machine parts. The modern SPD tools allow changing the surface properties of metals as is required and enable making the surface layers with the pre-set physical properties and optimal micro-relief. At the same time, the current SPD techniques are not rational as they need cumbersome equipment and consume much energy at inaccurate adjustment of treatment regimes. One of the ways of solving this problem is plastic deformation using vibro-impacts. The key advantages of this technology are accurate adjustability of hardening treatment regimes and applicability on-site. Moreover, the vibrio-impact technology can be assumed resource-saving.

The method of plastic deformation under vibro-impacts means multiple point-wise attacks on a part by means of a series of impacts with adjustable energy, speed of a treated part and feed of vibration–percussion machine. Different combinations of number and energy of impacts at the same energy input allow different rates of hardening.

At the present time the Department of the Transport Machine Building Technology and Operation at the Siberian State Transport University studies the use of electromagnetic percussive machines in handling various issues in construction, repair and reconditioning of railway lines [5–8]. For instance, vibration–percussion machines can: strengthen holes in rails, harden railway welds, make choke couplings, spikes etc., which requires a force impact on a treated material. The present paper authors think it is most expedient to strengthen alumothermic railway welds by vibration percussion when the jointed welds lose hardness as compared with the rail material. This study presents the experimental test data on influence of vibro-impact plastic hardening on microhardness and strengthened layer thickness of tread areas of railway weld joints.

For the experimentation on impact hardening of tread areas of railway welds, a laboratory impactor was designed to implement unit impacts with an energy T variable between 5 and 30 J. In order to find the minimum unit impact energy required to strengthen the welds and governing mass and size of an impacting tool, the tests used an indenter with a cylindrical flat face with a radius of 5 mm. After different energy unit impacts, the microhardness H and the strengthened layer thickness h were estimated in the zones of load application. The microhardness was estimated in accordance with the RF State Standard GOST 9450-76 on UIPMT-3 machine under magnification ×487. The load on the indenter was 200 g, the time under the load was 10 s. The microhardness was estimated on the surface of the tread area of the railway weld and depthward, at the depth step of 50 μm.

The indenter penetration depth is 32 μm at the maximum impact energy of 30 J. The microhardness grows from 3500 to 9000 MPa when the impact energy is changed from 5 to 30 J, and the strengthened layer thickness is increased from 0.6 to 2.4 mm (see the figure).

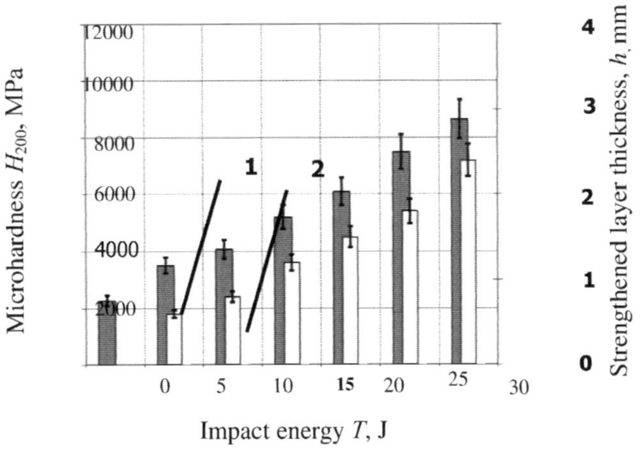

Effect of the impact energy on the microhardness 1 and strengthened layer depth 2.

In accord with the standard of alumothermic welding, the hardness of the tread areas of alumothermic railway welds is to be 380–400 NV, which conforms with the microhardness of 5800–6100 by Grigorovich. The experimental results have shown that the surface layer with the microchardness of 5800–6100 MPa is generated under impacting the tread area by the indenter with the cylindrical flat face with the radius of 5 mm at the minimum sufficient energy of unit blow of 20 J.

The application of the vibro-impact plastic deformation method of surface hardening allows recovering hardness of rail material in the zone of welding and reducing its crumple and spalling. Efficient implementation of the technology needs a mobile percussion installation with an electromagnetically driven impactor and a small capacity power supply (portable benzene power-generating plant).

References

[1] Afanasev AA 2007 *Improvement of Quality of Surface of Machine Parts* Belgorod (in Russian)
[2] Suslov AG 2003 Processing and enhancing efficienty of machine parts by plastic deformation *Inzh. Zh.* No 8 pp 8–12
[3] Balanovsky AE 2005 Surface stregthening of rails *Put Put. Khoz.* No 11 pp. 35–38
[4] Kirichek AV, Solovev DA and Lazutkin AG 2004 *Technology and Equipment for Static-and-Pulsed Surface Processing by Plastic Deformation* Moscow (in Russian)
[5] Kargin VA, Tikhomirova LB and Galay MS 2010 Hardenining of rail welds by surface deformation *Obr. Metal.: Nauch.-Tekh. Proizv. Zh.* No 3(48) pp 17–18
[6] Kargin VA, Tikhomirova LB, Galay MS and Kuznetsova YS 2015 Improving service propertis of welded joints produced by alumothermic welding *Welding International* Vol 29 No 3 pp 155–157
[7] Tikhomirova LB and Galay MS 2015 Increase in the hardness of alumothermic weleded joints of rails by surface plastic deformaiton *Polytransport Systems: Proc. VIII Int. Conf in the Framework of the EU–Russia Year of Science* Novosibirsk pp 564–567 (in Russian)
[8] Kargin VA, Tikhomirova LB, Abramov AD and Galkay MS 2010 Vibratuion-percussion deformation of wedled joints *Put Put. Khoz.* No 6 pp 22–23

Layout and variables of air-powered drive of small-size pneumatic winch

VP Gileta[1,2], AV Baris[2] and YuV Vanag[2]

[1]Chinakal Institute of Mining, Siberian Branch, Russian Academy of Sciences, Novosibirsk, Russia
[2]Novosibirsk State Technical University, Novosibirsk, Russia

E-mail: [1]pmx08@mail.ru

Abstract. The paper describes three layouts of a winch with a harmonic drive and a flexible gear in the form of a set of spindles: classical layout with the remote motor and barrel; layout with the harmonic drive and remote motor; layout with the barrel with the in-built drive members. The analytical models are developed to determine acting faces and lateral dimension of rigid gear teeth depending on the layout and on the design and kinematic variables of the winch and the drive.

It is preferred to lay service lines in towns using trenchless methods—piercing or puncturing when holes are made with partial or full compaction of soil. These methods are implemented with the help of Russian and foreign pneumatic percussion machines, for instance, pneumatically driven hammers M200, SO166, Typhoon, Grundomat [1].

In hole-making by puncturing, it is possible to drive steel pipes-casings with the shut front end in soil, or to make hole first with a pneumatically driven hammer and then to insert ceramic, plastic, asbestos or steel pipes in the hole.

Puncturing process has two stages: at the first stage a pipe–casing is driven in soil with the open front end, at the second stage soil core is removed from the pipe. Core removal is possible using different tools, including self-propelled suction heads with pneumatic percussion machines employed as impact impulse generators. Pulling-out of the suction head with soil core is executed by an auxiliary take-off assembly.

A feature of these technologies is the backblow of the pneumatic percussion machines, which is required to balance when in the course of hole-making. At early stages of hole-making, friction is low, and a force-feed gear is required to balance the backblow of a pneumatic percussion machine. To this effect, it is possible to use startings in the form of a rod power-units or a leverage mechanism; the force is generated by an operator in this case [4]. Since startings are not in serial production, usually leverage mechanisms or winches are chosen.

Winches are used to pull suction heads with soil core from pipes. As experience shows a winch can balance the backblow of a pneumatic percussion machine at early stages of pipe driving. The current winches have hydraulic, pneumatic, electric and manual drives [6]. Dimension and weight of drives grow in proportion to the value of the pulling force or carrying capacity of a drum (and number of drums). For the confined environment of hole-making with pneumatic percussion machines, it is preferable to chose winches with pneumatic and manual drives.

Inasmuch as the current winches are big and heavy, while manual winches provide low pulling force, it is required to design small-size winches. The aim of this study is designing a winch with a pneumatic drive, of small size and low weight, to balance the backblow of the pneumatic percussion machine at early stages of hole-making or to work in a confined area.

The main components of a winch are motor, reducing gear and rum. This study focuses on the design of a small-size winch with a gear system based on a harmonic gear in the form of a set of spindles, of small size and with a wide-range transmission ratio.

The harmonic gear with spindles consists of generator 1, internal rigid gear 2, disc guide 3 and spindle 4 (Figure 1).

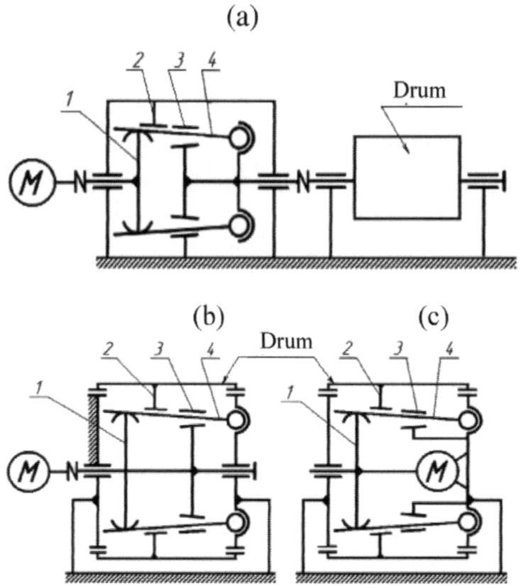

Figure 1. Variants of layout of a winch with a chain harmonic gear. Explanation is in the text.

There may be three variants of layout of a winch with a harmonic gear: outboard motor and drum—classical layout (Figure 1a), which allows collapsible winch to facilitate the winch transport to a work site; outboard motor and harmonic gear integrally mounted in a drum (Figure 1b); all elements of drive, including motor, are embedded in a drum (Figure 1c). Variants (a) and (b) in Figure 1 allow the smallest vertical and horizontal dimensions of a winch.

Capabilities of a small-size winch with a harmonic gear with spindles and geometric and kinematic parameters of the gear depend on dimension of the rigid gear. The most important stage of the rigid gear design is shaping of acting faces of the gear teeth. There are many methods of gear tooth shaping: Gokhman, Olivier, kinematic method. The most popular method is based on the classical apparatus of differential geometry [5–8].

To define acting faces of rigid gear teeth, it is necessary to know trajectory of spindle axes in the coordinate system XYZ connected with the rigid gear.

Let spindle axes move along a trajectory described by a parametric equation [9]

$$\begin{cases} x_M = f_x(t, L_m), \\ y_M = f_y(t, L_m), \\ z_M = f_z(t, L_m), \end{cases} \qquad (1)$$

where x_M, y_M, z_M are the coordinates of the point M belonging to the surface (1) and concurrently to a spindle axis; t is the time of travel of the axis along the chain; L_m is the distance between the point M and the center of the spherical base of a spindle.

The coordinates of the point P belonging to the acting face of a tooth of the rigid gear can be presented y the coordinates of a radius-vector

$$\overrightarrow{OP} = \overrightarrow{OM} + \overrightarrow{MP} \tag{2}$$

where the vector $\overrightarrow{MP}$ is given by

$$\overrightarrow{MP} = \pm \ \vec{n} \cdot r, \tag{3}$$

where $\vec{n}$ is the unit normal vector drawn to the surface (1); r is the radius of the acting face of a spindle. The vector $\vec{n}$ is obtained by normalizing of the vector $\vec{N}$ determined by a vector product of two non-collinear vectors tangential at the point M to the trajectory of the spindle axis

$$\vec{N} = \vec{\mathrm{v}} \times \overrightarrow{MW} \ . \tag{4}$$

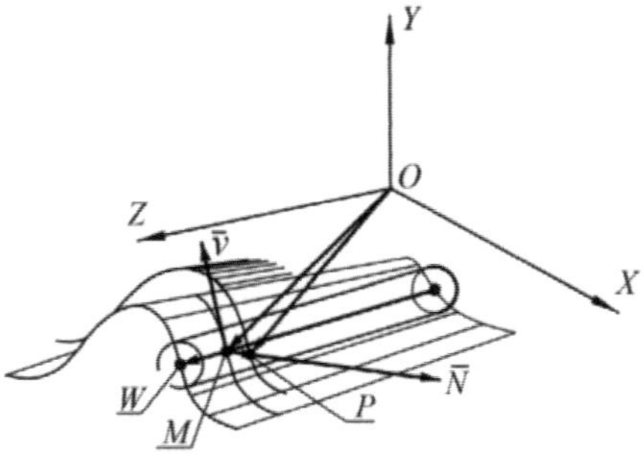

Figure 2. Schematic to determine the acting face of a rigid gear spindle.

In our case, the vector $\bar{v}$ is the velocity vector of the point M

$$\begin{cases} v_x = \dfrac{\partial f_x(t, L_m)}{\partial t} \\ v_y = \dfrac{\partial f_y(t, L_m)}{\partial t}, \\ v_z = \dfrac{\partial f_z(t, L_m)}{\partial t} \end{cases} \tag{5}$$

where f_x, f_y, f_z are the functions of the coordinates of the point M belonging to the spindle axis, in the coordinate system connected with the rigid gear.

Another vector of the product (4) is the vector $\overrightarrow{MW}$ belonging to the surface (1) and coinciding with its generating line

$$\begin{cases} x_{\overrightarrow{MW}} = x_W - x_M, \\ y_{\overrightarrow{MW}} = y_W - y_M, \\ z_{\overrightarrow{MW}} = z_W - z_M. \end{cases} \tag{6}$$

We use the expressions (1)–(6) to derive the equations for the acting faces of the rigid gear teeth in the parametric form

$$\begin{cases} x_P = \psi_x(t, L_m), \\ y_P = \psi_y(t, L_m), \\ z_P = \psi_z(t, L_m), \end{cases} \tag{7}$$

where x_P, y_P, z_P are the coordinates of the point P. The values t and L_m are the curvilinear coordinates of the point. The relations (7) are the closed spatial curves with the fixed parameter L_m and the variable parameter t. The surface described by (7) can be represented as a population of such lines for the variable values of L_m if L_m is constant, for the variable t we obtain a curved line that is the line of contact of a spindle and a rigid gear tooth.

Figure 3 shows the surface of the rigid gear teeth. The equidistant curve for the gear teeth is plotted out of the spindle axis travel trajectory.

The acting faces of the rigid gear teeth are to be limited by the design parameters of the harmonic gear in the direction of its axis coinciding with the coordinate Z of the chosen coordinate system *XYZ*.

The acting faces of the rigid gear teeth are obtained by cutting out the surfaces, described by the relations (6), by the planes Π_1 and Π_2, which are parallel to the plane XY at the distances L_{max} and L_{min} (Figure 4).

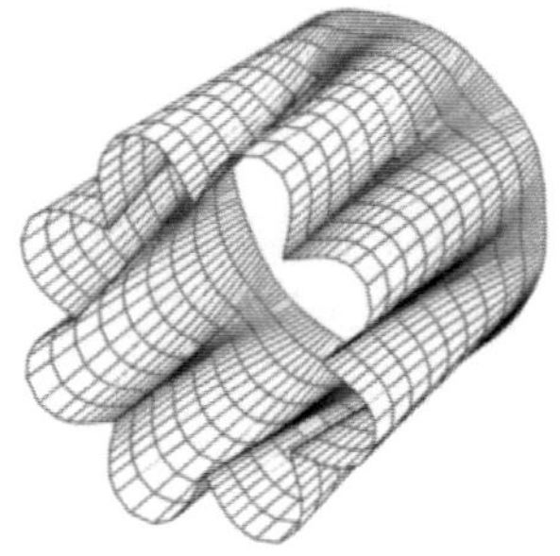

Figure 3. Surface of a rigid gear tooth.

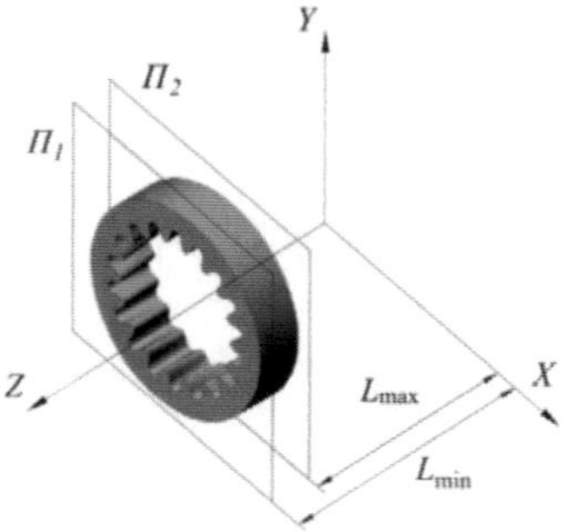

Figure 4. Rigid gear. Explanation is in the text.

Conclusion

The authors have offered three layouts of a winch with a harmonic gear with spindles: outboard motor and drum, outboard motor and harmonic gear integrally mounted in a drum and with a drum with all elements of the gear in-built.

The developed model of toothing of the harmonic gear and the rigid gear enables synthesizing the toothing at the linear contact of spindles.

The limiting conditions for the longitudinal parameters of the rigid gear teeth are determined as function of layout and design of a winch with a harmonic gear with spindles.

References

[1] Kostylev AD, Gileta VP et al 1987 *Pneumatically Driven Hammers in Construction* Novosibirsk: Nauka (in Russian)

[2] Smolyanitsky BN, Tishchenko IV, Chervov VV, Gileta VP and Vanag YuV 2008 Sources of productivity gain in vibro-impact driving of steel eleents in soil in special construction technologies *Journal of Mining Science* Vol 44 No 5 pp 490–496

[3] Gileta VP, Vanag YuV, Tishchenko IV, Vedel PV and Volkov NS 2015 Application of a small-size pneumatic percussion mechanism *Aktual. Problm. Mashinostr.* No 2

[4] Kershenbaum NYa and Minaev VI 1984 *Horeizontal and Vertical Percussive Hole-Making* Moscow: Nedra (in Russian)

[5] Artobolevsky II 1988 *Theory of Mechanisms and Machines: University Textbook* Moscow: Nauka

[6] Vildgaber E *1948 Pirinciples of Toothing in Conical and Hypoid Gears* Moscow: Mashgiz (in Russian)

[7] Litvin FL 1960 *Theory of Toothing* Moscow: Gos. Izd. Fiz.-Mat. Lit. (in Russian)

[8] Frolov KV *Theory of Mechanisms and Machines: University Textbook* Moscow: Vyssh. Shkola (in Russian)

[9] Baris AV and Gileta VP 2014 Chain harmonic gear *Izv. vuzov. Stroit.* No 5

Design of abrasive tool for high-rate grinding

AS Ilinykh

Siberian State Transport University, Novosibirsk, Russia

E-mail: asi@stu.ru

Abstract. The experimental studies aimed to design heavy-duty abrasive wheels for high-rate grinding are presented. The design of abrasive wheels with the working speed up to 100 m/s is based on the selection of optimized material composition and manufacture technology of the wheels.

1. Introduction

Abrasive machining on wheels with a speed up to 50 m/s is currently used in many branches of industry. It has been already proved that high speed of abrasive wheels can enhance efficiency of the treatment and reduce the force and temperature impact exerted on a treated material [1, 2]. Abrasion technologies increasingly use wheel speeds up to 100 m/s and higher.

Under high-force and high-speed abrasion, a wheel experiences high mechanical and thermal stresses that can induce rapid wear and even damage of the abrasive tool.

For this reason, one of the main constraints to application of new high-capacity technologies of material processing is the absence of an abrasive tool possessing increased rupture strength [3].

The known ways of enhancing the strength of an abrasive wheel include:

—improvement of wheel design by a reinforced nonworking surface;

—application of reinforcing elements;

—manufacture of a wheel of maximally close texture and uniform structure by hot shaping;

—optimized composition of a binder, including modifiers, to increase adhesion between the binder and inorganic components.

The aim of these studies was the design of an abrasive wheel capable to operate at high speeds up to 100 m/s and to ensure the operation safety owing to optimized mix formulation and technology of abrasive wheel manufacture.

2. Test procedure

In order to design an abrasive tool, it is required to know its strength characteristics. This is particularly important to ensure safe operation of high-speed abrasive wheels.

The dependence of the breaking speed V_{break} of abrasive wheel on its strength, density and geometry is known [4]

$$V_{\text{break}} = \sqrt{\frac{g \cdot \sigma_{\text{T}}}{\gamma}} \cdot \sqrt{\frac{3}{1+\alpha+\alpha^2}}, \tag{1}$$

where g is the acceleration of gravity; σ_{T} is the ultimate tensile strength; γ is the density of material the wheel is made of; α is the shape factor of the wheel (inside diameter/outside diameter).

According to the data obtained in the tests of samples and wheels with bakelite resin binder subjected to tension, flexure and compression to determine breaking speed under rotation [5], the most significant strength characteristic of the material an abrasive wheel is made of is the flexing strength. In this case, the relation (1) can be given by

$$V_{\text{break}} = K\sqrt{\frac{\sigma_{\text{F}}}{\gamma(1+\alpha+\alpha^2)}}, \qquad (2)$$

where K is an empirical parameter; σ_S is the static flexing strength.

It is proved experimentally that $K = 9.4$ and $K = 11.3$ are valid for wheels without and with reinforcing elements, respectively. So, with the known K, σ_F, γ and α, it is possible to calculate the estimated breaking speed of an abrasive wheel. To determine physical properties, samples in the form of bars 150×90×20 mm were manufactured and tested.

In the static flexure tests, the destructive stress was determined using the formula

$$\sigma_{\text{F}} = \frac{3FL_0}{2BH^2}, \qquad (3)$$

where F is the load at the moment of failure of a sample; L_0 is the spacing of bases; B is the width of a sample; H is the height of a sample, mm.

The flexure testing machine P-10 ensured uniform velocity of the load head and the bases (100 ± 20 mm/min) at the measurement error of ± 1%. Each mix composition of samples was tested three times, and the average flexing stress was assumed as the test result.

The hardness of the samples was determined using sand blast model 910 in accord with the Russian State Standard GOST 18118.

The density γ was found from the dimensions and weights of the samples.

2. Test results

There were three variants of mix composition. Five samples of each mix composition were manufactured and tested after bakelization.

Variant 1: zirconium aluminum oxide (AO) bars with the grain size of 125, the limit content of grains $V_G = 54\%$ of the sample volume, content of binder $V_B = 34\%$, content of pores $V_P = 12\%$; the mix was not conditioned.

Variant 2: AO bars with grain sizes 160, 125 and 100 at a ratio of 40–40–20%, $V_G = 58\%$, $V_B = 32\%$, $V_P = 10\%$; %; the mix was not conditioned.

Variant 3: AO bars with grain sizes 160, 125 and 100 at a ratio of 40–40–20%, $V_G = 58\%$, $V_B = 38\%$, $V_P = 4\%$; %; the mix was conditioned under the temperature of 60°C for 8 hours.

The test results are reported in Table 1.

Table 1. Physical properties of the tested samples.

Variant no.	Structure, %			Density, g/cm^3	Flexing strength, Pa
	$V_З$	V_C	$V_П$		
1	54	34	12	2.93	371
2	58	32	10	3.31	405
3	58	38	4	3.59	448

The strength of *Variant 1* bars is higher than the strength of *Variant 2* bars by 9%, and conditioning of the mixture before shaping (*Variant 3*) increases the flexing strength by 20%.

Also, the influence of binders and basic materials on heat stability of the abrasive–bakelite mixtures was analyzed. The heat stability was determined by estimating flexing strengths of bakelized bars before and after heat exposure (250 ± 5)°C for 18 hours.

Three mixtures of the same structure samples ($V_G = 54$, $V_B = 40\%$, $V_P = 6\%$ were tested.

Variant 1: powdered binder—pulverized bakelite (SFP-011A), unmodified; basic material—pyrite.

Variant 2: powdered binder—pulverized bakelite modified by polyvinyl buteral (SFP-0129A), unmodified; basic material—pyrite.

Variant 3: a mix of powdered basic materials of pyrite, copper power and micro-powder of silicon carbide and pulverized bakelite modified in accordance with *Variant 2*.

The test data are given in Table 2.

Table 2. Heat stability of the tested samples.

Description	Unit	Mixture variant		
		1	2	3
Strength before thermal treatment	Pa	430.6	428.3	457.9
Strength after thermal treatment	Pa	343.4	357.1	390.1
Strength variation after thermal treatment	%	–20.3	–16.6	–14.8

These data show that the maximal heat stability (i.e. minimum reduction in strength) is a feature of the samples manufactured using modified pulverized bakelite and compound basic material.

Thus, the mixture and manufacture method, which is the use of zirconium aluminum oxide with the grain size composition of 160, 125 and 100 in combination with the modified binder and the compound basic material with conditioning, ensured improved physical properties of the abrasive–bakelite tool and were taken as the model for the further research.

High-density abrasive bakelized tools were manufactured using hot shaping under low and high temperature from 60 to 100°C and from 170 to 180°C, respectively. Under low-temperature hot shaping, the organic binder was softened, the mixture became yielding and was well compacted under pressure. Under further increase in the temperature up to the high-temperature hot shaping range, particle polymerization of the binder took place (hardening) and almost nonporous material was manufactured.

The low- and high-temperature hot shaping tests used samples made of the mixture composed of AO with the grain size 160, 135 and 200 to make the active face of an abrasive wheel and the mixture composed of silicon carbide with the grain size 12N to make the nonworking face of the wheel.

The data of these tests are compiled in Table 3.

Table 3. Physical properties of the test samples.

Parameter	Unit	Low-temperature hot shaping		High-temperature hot shaping	
		Acting face	Nonworking face	Acting face	Nonworking face
Density	g/cm^3	3.30	2.41	3.38	2.69
Rupture stress under flexure	Pa	760 ± 50	800 ± 40	820 ± 40	1070 ± 50
Ultrasound speed (USS)	m/s	4620 ±2 0	4260 ± 40	4800 ± 30	5010 ± 50
Cratering depth	mm	0.25 ± 0.1	1.8 ± 0.15	0.20 ± 0.1	1.6 ± 0.15
Hardness degree	–	EH	EH	EH	EH
Estimated breaking speed	m/s	145 ± 5	174 ± 5	140 ± 4	190 ± 5
Estimated strength margin	–	2.1 ± 0.15	3.05 ± 0.15	2.2 ± 0.1	3.6 ± 0.2

EH—extremely hard

It is clear from Table 3 that the bars manufactured under high-temperature hot shaping possess higher density, strength, USS and hardness than the bars made using the low-temperature hot shaping technique.

3. Conclusion

The author has accomplished the experimental research aimed to design high-speed (100 m/s) abrasive wheels. The physical properties of an abrasive bakelized composite studied depending on mixture composition and manufacture technology allowed optimization of the both factors, and a pilot run of abrasive wheels 25-250×50×76 38A 125 EH (extremely hard) B (bakelized) R (reinforced) 100 m/s of the standards geometry and strength has been manufactured.

Structurally, the abrasive wheels have the acting faces made of zirconium aluminum oxide and the nonworking faces made of fine-grained AO or silicon carbide, which allows enhanced strength of the wheel at the lower cost due to saving of expensive AO.

To enhance operation safety, the wheels are reinforced with the four layers of roving along the outside diameter as against the two-layer roving of series-produced abrasive wheels for operation at the speed of 50 m/s.

The samples made with the optimized mixture parameters and manufacture technique have been subjected to strength testing.

A pilot run of 10 abrasive wheels was manufactured using the high-temperature hot shaping method and: $V_G = 54$, $V_B = 44\%$, $V_P = 2\%$ to make the acting face and $V_G = 50$, $V_B = 48\%$, $V_P = 2\%$ to make the nonworking face; the mixtures before shaping were conditioned under 65 ± 5°C in the course of 12 hours; shaping press had a force of 250 t, the hot shaping temperature was 180 ± 5°C, holding time under pressure was 40 min, bakelization under 185°C took 16 hours; after bakelization, the wheels were machined with a diamond tool along the outside diameter and on faces.

To determine the actual strength margin, 2 wheels from that pilot run were subjected to rotation until rupture on test bench SBIP-350. The first wheel was ruptured at 12230 rpm (160 m/s), the second wheel—at 13030 rpm (170 m/s); thus the strength margin made 2.6–2.9.

The rest 8 wheels were set aside for the full-scale testing in order to specify their operating factors.

Finally, the designed abrasive wheels 25-250×50×76 38A 125 EH B R 100 m/s) conform with the assignment for their development. The new wheels have the size, disbalance, safety factor and operating factors in accordance with the initial specifications.

References

[1] Zubarev Yu.M. and Priemyshev AV 2010 *Theory and Practice of Enhancement of Abrasion Treatment Effciency* Saint-Petersburg (in Russian)

[2] Orobinsky VM 2000 *Abrasion Treatment Methods and Their Optimization* Mir (in Russian)

[3] Pilinsky AV 2015 Innovations and challenges in high-speed and ultra high-speed abrasion treatment *Vektor Nauki TGU* No 2(32–2) pp 136–143

[4] Bakul VN 1975 *Design Principles and Manufacture Technology of Abrasive and Dimanond Tools* Moscow (in Russian)

[5] Netrebko VP and Korotkoe AN 1992 *Strength of Abrasive Wheels* Moscow (in Russian)

IOP Conf. Series: Earth and Environmental Science **53** (2017) 012025 doi:10.1088/1755-1315/53/1/012025

Determination of variables for air distribution system with elastic valve for down-the-hole pneumatic hammer

AYu Primychkin[*]**, AS Kondratenko and VV Timonin**

Chinakal Institute of Mining, Siberian Branch, Russian Academy of Sciences, Novosibirsk, Russia

E-mail: [*]808@nn.ru

Abstract. The air distribution system of down-the-hole pneumatic hammer 105 mm in diameter is updated to enhance drilling efficiency. The design model of the down-the-hole pneumatic hammer is constructed in ITI SimulationX environment. The basic variables of the air distribution system with an elastic valve are determined so that to ensure increased impact energy at the limited pre-impact velocity and the same machine size.

The modified design of a pneumatic hammer with a diameter of 105 mm is intended to enhance efficiency of drilling [1]. Such diameter machines are used in underground blast hole drilling with rock drill model NKR-100. The most widely employed currently is pneumatic hammer model P105PM designed at the Institute of Mining, SB RAS, based on air hammer M48 and manufactured by many plants, including Stary Oskol Machine Shop.

The new modification of a pneumatic hammer is based on the layout of PP105-EN hammer [3], with an air distribution system equipped with an elastic valve, oriented by PP105-K hammer designed by V. Gaun [4]. The present R&D was aimed to increase blow energy of an air hammer at the limited pre-blow velocity and the preserved dimensions of the machine. The earlier research has shown that pneumatic hammers with elastic valves exhibit higher capacity as against valveless machines of the same sizes. Figure 1 shows the layout of PP105-EN hammer. An elastic ring-shaped valve (ERV) is installed in an adapting pipe 1 to control inlet of compressed air from the main line to the power stroke chamber B. Also from the adapting pipe, compressed air is fed to a central pipe 2 for bottom hole purge. The side holes 3 in the central pipe perform continuous feed of a cell C to supply the back stroke chamber A or the power stroke chamber B depending on the position of the striking part 6. ERV closure takes place when the power stroke chamber drops due to the opening of the exhaust channels 4 made in the walls of the housing 9.

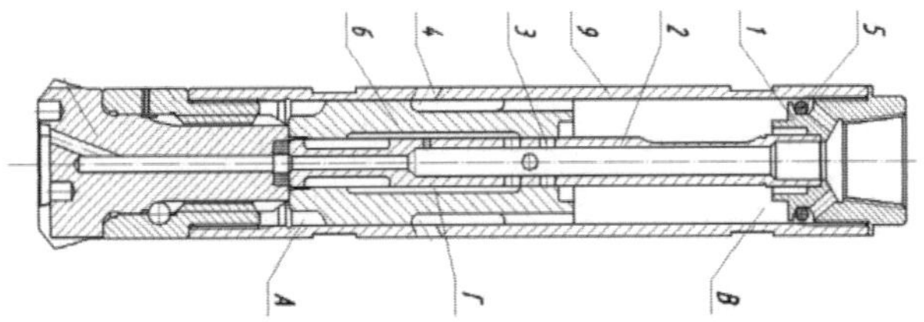

Figure 1. Down-the-hole pneumatic hammer PP105-EN (legend is explained in the text).

The feature of the elastic valve is that it is round. Such valve prevents from variation in the valve passage pressure p^* [5]. According to the research findings [6], it is possible to vary the average pressure above the valve and to adjust ERV actuation pressure by placing a rectangular cross-section valve with a face on the side of the air flow exit.

The given study task was to determine rational sizes of chambers, areas of channels, dimension and weight of striking part, positions of distributing edges of the striking part and, certainly, parameters of ERV with a view to ensuing maximum possible energy and capacity of hammering.

A quick and economical way of development of air distribution systems for pneumatic percussive machines is the computer-aided modeling. This method allows experimention without manufacture of a full-scale sample of the machine.

The down-the-hole pneumatic hammer PP105-EN with an elastic valve installed in the system of air distribution was modeled in ITI SimulationX. The model included the key parameters of the machine: volumes of the working chambers, areas of channels, times of opening and closure of the channels, weight and areas of the striking part on the sides of the power stroke and back stroke chambers. Also, the model contained the earlier modeled elastic valve [7], that described the operating cycle of the valve, namely, control of the flow area of the passage feeding the power stroke chamber depending on the valve pressure difference. The model of the valve accounted for the valve stiffness, pre-tension, maximum travel, elasticity of the valve material and the length of the face (if present).

This model of a pneumatic hammer enables variation of any design paraneters of an air distribution system and allows controlling the associated change in the energy characteristics of the hammer: *A*—blow energy; *v*—blow frequency; *W*–imact capacity; *u*—pre-blow velocity of the hammer head; *q*—compressed air flow rare, and others.

The ITI SimulationX model of the down-the-hole pneumatic hammer PP105-EN is depicted in Figure. 2. The elastic valve is represented by a throttle 2 controlled by a function block 4. In the model, the valve acutation pressure was calculated using a simplified option as this is sufficient for a design calculation and there is no need to make detailed descriptuon of the valve travel [5].

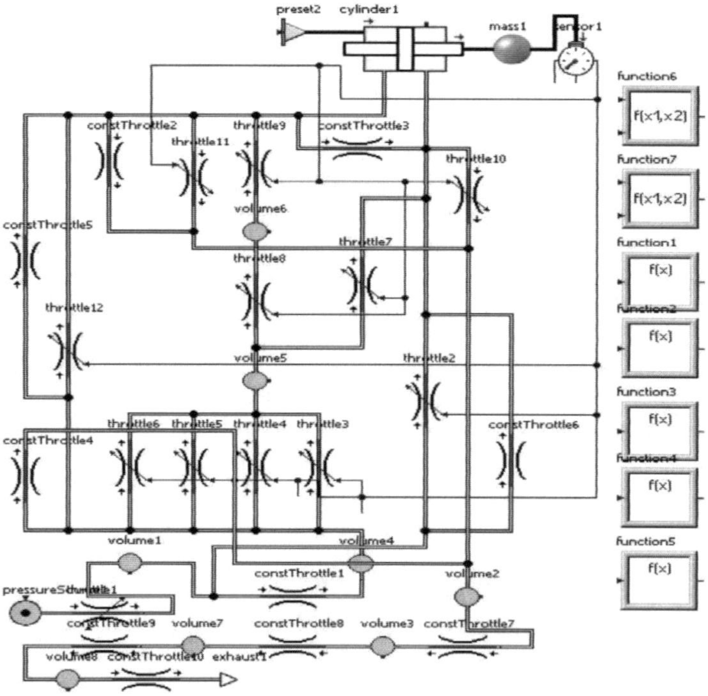

Figure 2. Pneumatic hammer PP105-EN model in ITI SimulationX.

The modeling involved various combinations of the variables of the air distribution system. The back stroke chamber volume (vol0A) was varied from 0.1 to 0.279 l, and the power stroke chamber volume (vol0B)—from 0.05 to 0.12 l. The striking part weight m was ranged between 4.08 and 4.62 kg, the elastic valve capacity (throttle2) was varied from 36 to 48 l/(bar·s), the edge of the exhaust hole (throttle10) was displaced by 0 to 26 mm (relative to the forward position of the hammer head), the pipe edge (throttle7) was shifted by 70–130 mm. The main feed line pressure was assumed 0.5 or 0.6 MPa. The maximum pre-blow velocity was limited to 9 m/s [8]. The overall calculation variants totaled 33.

After the calculations, we selected design variables to ensure the highest blow performance of the percussion machine at the minimized flow rate of compressed air. The plots of the operational cycle of the pneumatic hammer with the selected variables are given in Figure 3.

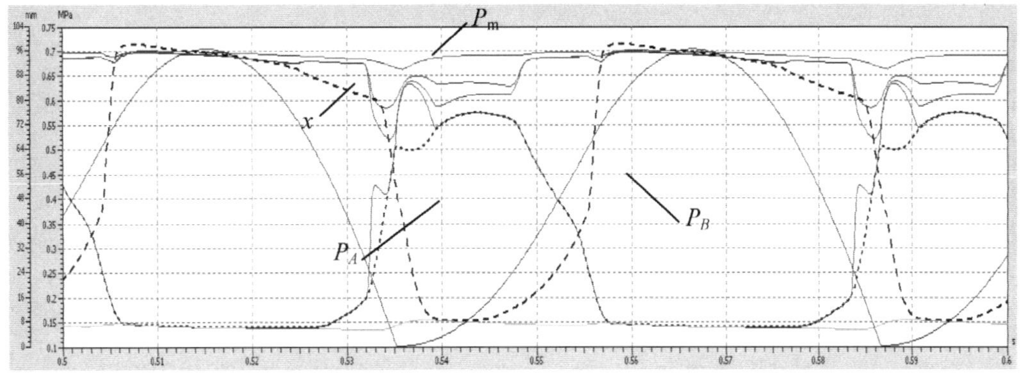

Figure 3. Operational cycle of pneumatic hammer with the rationally adjusted air distribution system: x—striking part travel; p_A—pressure in the back strike chamber A; p_B—pressure in the back strike chamber B; P_m—pressure in the main line.

The rational design of the air distribution system involved such finalized variables as: back stroke chamber volume (vol0A)—0.279 l; power stroke chamber volume (vol0B)—0.05 l; striking part weight m—4.42 kg; capacity of the elastic valve (throttle2)—48 l/(bar·s); position of the exhaust hole edge (throttle10)—0 mm (relative to the forward position of the hammer head); position of the edge on the pipe (throttle7)—70 mm. The main feed line pressure was 0.6 MPa.

With the rationally set air distribution system, the pneumatic hammer performance is: blow energy A = 162 J; blow frequency v = 19.5 s^{-1}, impact capacity W = 3.2 kW; air flow rate q = 0.036 $m^3/(kW \cdot s)$; pre-blow velocity u = 8.6 m/s. As compared with the series production air hammer P105-PM, the new design hammer has the higher blow energy by 40%, the increased impact capacity by 18%, the lower air flow rate by 20% and the reduced blow frequency by 19%.

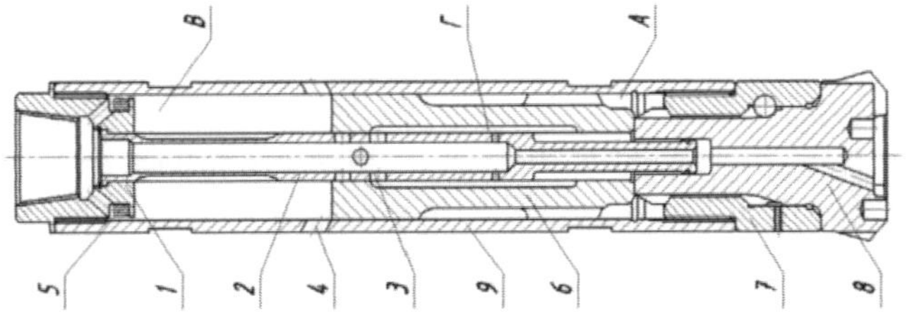

Figure 4. Modernized down-the-hole pneumatic hammer PP105-1KT. Description is in the text.

On the ground of the computer modeling results, the layout of the pneumatic hammer PP1-5-EN was modified: the central pipe 2 was elongated; the valve 6 was re-shaped and the elastic valve 5 of rectangular cross-section and with a face was included in the layout. The modernized pneumatic hammer is intended for drilling holes with a diameter of 105 mm. The modified layout is shown in Figure 4. The mode of operation of this hammer is the same as with PP105-EN (refer to Figure 1).

Comparing the performance of the series production machines and the modernized pneumatic hammer design, the latter exhibits the considerable increment in the blow energy and capacity at the low air flow rate.

The new-design pneumatic hammer was trialed at a commercial scale in Gubkin Mine, KMAruda, between 12 and 18 June 2016.

Drilling using rock drills NKR-100MPA was carried out in quartzite with a low content of metals, with the strength up to 200 MPa. Drilling of blast holes was oriented top downward. Compressed air pressure in the main line of the pneumatic hammer was 0.7 MPa.

The commercial trials showed that penetration rate is 1.5–2 times higher with the modernized pneumatic hammer than with the series production hammer P105-PM.

References

[1] Eremenko VA, Karpov VN, Timonin VV, Barnov NG and Shakhtorin IO 2015 Basic trends in development of drilling equipment for ore mining with block caving method *Journal of Mining Science* Vol 51 No 6 pp 1113–1125

[2] Esin NN 1976 *Down-the-Hole Pneumatic Percussion Machines for Hole Drilling* Novosibirsk: Nauka (in Russian)

[3] Timonin VV 2015 Downhole air hammers for underground mining *Gorn. Oborud. Elektromekh.* No 2 pp 13–17

[4] Gaun VA 1981 Author's Certificate No. 848615 Pneumatic percussive device *Byull. Izobret.* No 27

[5] Petreev AM and Primychkin AYu 2015 Influence of air distribution system on energy efficiency of pneumatic percussion unit of circular impact machine *Journal of Mining Science* Vol 51 No 3 pp 562–567

[6] Petreev AM and Primychkin AYu 2016 Ring-type elastic valve operation in air hammer drive *Journal of Mining Science* Vol 52 No 1 pp 135–145

[7] Petreev AM, Voronstov DS and Primychkin AYu 2010 Ring-shaped elastic valves in the air percussion machines *Journal of Mining Science* Vol 46 No 4 pp 416–424

[8] Serpeninov BN and Nikonova IP 1976 *Transmission of Axial Impact in Hammer Head–Rod–Rock System as Regards Some Percussive Machines: Report* Novosibirsk: IGD SO RAN (in Russian)

Straight hole drilling machines for coal mines

VV Timonin*, DI Kokoulin, SE Alekseev and B Kubanychbek

Chinakal Institute of Mining, Siberian Branch, Russian Academy of Sciences, Novosibirsk, Russia

E-mail: *timonin@misd.ru

Abstract. The authors prove the demand for drilling machines capable of making long straight holes in rocks with the strength up to 120 MPa. The paper describes the designed, manufactured and tested down-the-hole hammers for rotary–percussion drilling of long straight directional holes. The hammers have been delivered to Berezovskaya Mine for further trial and commercial operation.

Labor content and costs of underground coal mining are to a large degree conditioned by methane drainage, service and exploration hole drilling. Total hole drilling in mines in Russia and the CIS countries makes tens thousands kilometers. For example, Karaganda Department of ArselorMittal Timirtau annually drills more than 300 km of underground holes [1]. Drilling equipment used in the coal industry in Russia is both of domestic and foreign manufacture. Unfortunately, the capacity of these drilling tools falls short of the current technologies in coal mining. Actual operation of this equipment has revealed their common drawbacks, namely:

—limited application in hole drilling in rocks with the strength of 120 MPa;

—complicated drilling of straight long holes both in coal and in rocks.

As for the first drawback, rotary drilling equipment used in coal mines is efficient in rocks with the strength not higher than 80 MPa. As regards the second drawback, in the course of rotary drilling in coal and in rocks using drill bits and rolling cutter bits, essential axial forces and torsion are applied on the drill stem from the side of the rock drilling machine; these forces induce lateral forces, and a drill hole deviates from the pre-assigned direction as a result [2].

From the experience of drilling in coal mines, it is possible to drill straight holes by varying drilling rate by changing rotary speed and feed speed of drill stem. However, in spite of apparent operating efficiency, this approach also has limitations. First of all, straight hole drilling is ensured by reduction in feed speed, i.e. drilling penetration rate, which increases nonproductive rig time; second, this method of drilling is ineffective in hard rocks. So, it is required to design drilling machines capable of drilling straight holes either in coal or in rocks of high hardness.

The known method of drilling almost straight holes is rotary–percussion drilling by down-the-hole hammers. In this case, rock is destructed by the hammer head directly impacting the hole bottom. As compared to other techniques, this method allows minimum pull-down pressure from drilling tool to hole bottom, low speed of rotation and minor torque applied to the drill stem. This makes it possible to minimize distortion of the drill stem and to drill almost straight holes.

Deloading of a drill stem is feasible with down-the-hole hammers of minimum diameter relative to the given drilling machine. In this case, forces induced by the impact interaction between the hammer and the hole bottom are applied closer to the hole bottom center, which reduces torsion transferred to the drill stem.

Direct hole drilling requires uniform impact loading of hole bottom. This is in a greater degree achieved under central-force collision of hammer head and striking part when impact load is more evenly applied on them. Central-force collision enables reduction in transverse vibration of the hammer, which improves straightness of hole drilling.

Also, the experience of drill bit engineering in Russia and abroad has shown that higher accuracy of drilling is achieved with drill bits making longer coring, which ensures better fixturing of the drill bit relative to the hole axis [3].

Based on these considerations, the Institute of Mining, SB RAS, has designed a small-size drill hammer PNB76 (Figure 1) [4] with drill bit KNB76 to ensure coring such that to center the drill bit relative to the hole bottom (Figure 2).

Design of an air hammer for small diameter hole drilling is rather challenging for the drill hole diameter imposes strict constraints on the cross dimension of DTH hammers and limits potential of the wanted-size active areas of a hammer head. Aiming to enlarge active area of a hammer head, the proposed design of DTH hammer provides that the hammer head area subjected to pressure is more than the cross section area of the cavity of the hammer housing. The hammer is of closed type, which ensures higher wear resistance of friction bearing backs and improves quality of bottom hole cleaning from drill cuttings not to be overground.

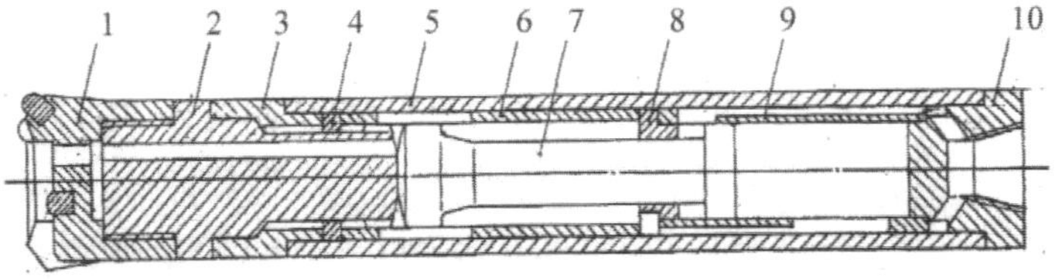

Figure 1. Layout of small-size DTH hammer PNB76: 1—removable drill bit; 2—plunger; 3—bearing box; 4—front split collar; 5—housing; 6—front quill shaft; 7—striking part; 8—rear spit collar; 9—rear quill shaft; 10—adapater connector.

Figure 2. Drill bit KNB76 of TH hammer PNB76.

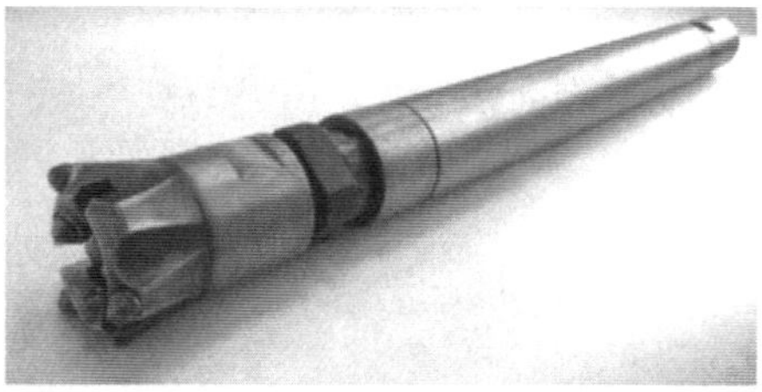

Figure 3. Trial model of the designed hammer.

The central-force collision of the striking part and the drill bit improves blow energy transmission due to reduction in cross vibration and relieves stresses in the colliding parts owing to more uniform distribution of impact load in cross sections of the colliding parts. Valveless air distribution and simple design ensues fault-free performance of the hammer. In the proposed layout, air is only exhausted from the power stroke chamber, and there is no exhaust from the idle run chamber. "Spurious" volume non-productively filled by compressed air is diminished in this case, which cuts down air consumption and makes the machine cost-effective.

A trial model of the down-the-hole hammer has been manufactured by Spetsgidravlika (Figure 3).

DTH hammer PNB76 has been tested on a brassboard (Figure 4).

Figure 4. Full-scale testing of DTH hammer PNB76.

Characteristics of the hammer recorded during the tests under compressed-air pressure of 0.6 MPa are shown in Figure 5.

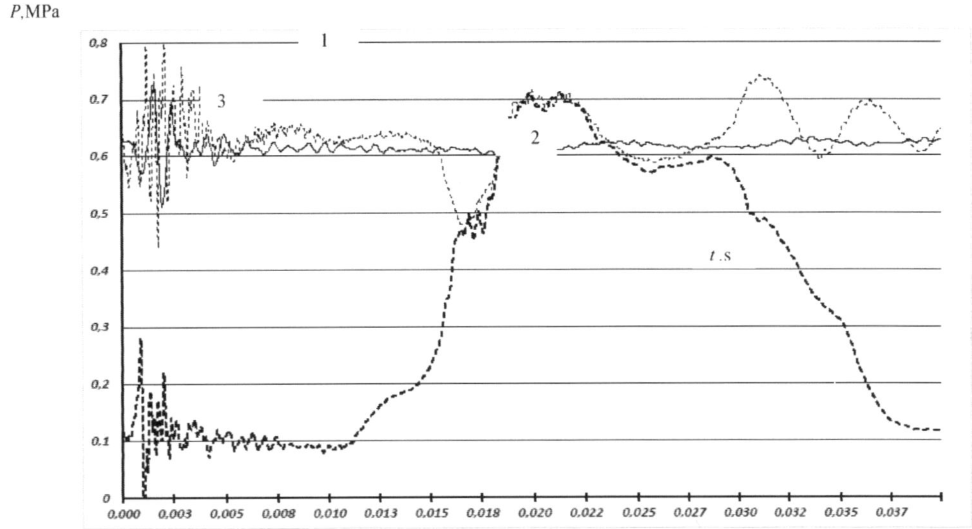

Figure 5. Pressure in the chambers of DTH hammer PNB76: 1—in the main line; 2—in the power stroke chamber; 3—in idle stroke chamber.

After processing of the curves using the known procedure [5], performance of PMB76 hammer has been determined:

Drill hole diameter	76 mm
Hammer housing diameter	63.5 mm
Driving pressure	0.6 MPa
Blow energy	40 J
Blow frequency	1500 min^{-1}
Hammer head weight	1.45 kg

Based on the positive test data, the Institute of Mining has proposed Berezovskaya Mine to continue jointly R&D on straight hole drilling in a production environment.

On request of Berezovskaya Mine, Spetsgidravlika has manufactured three DTH hammers PNB76 with drill bits KNB76 for operation of drill machine SBR400 designed and patented by the Institute of Mining, SB RAS.

At the present time, the equipment has been delivered to the customer. It is intended that the hammers will for the first turn been used to drill straight service holes more than 130 m long in rocks with the strength of 120 MPa.

References

[1] Kashtanov KS, Polchin AI, Udovov DB, and Butler N 2010 Integrated approach to draining of gases at the Coal Department of ArselorMittal Timirtau *Ugol* No 1 pp 31–34

[2] Repin AA, Kokoulin DI, Alekseev SE, Karpov VN, and Shakhtorin IO 2014 Small-size air hammer for directional long hole drilling in coal mines *Prospects for Innovative Development in Coal Mining Regions of Russia: IV International Conference Proceedings* Prokopievsk (in Russian)

[3] Repin AA, Alekseev SE and Kokoulin DI 2012 Long directional hole drilling *Resource-Saving Technologies in Blasting: Conference Proceedings* Ekaterinburg (in Russian)

[4] Repin AA, Alekseev SE and Karpov VN 2012 RF Patent No. 121854. Down-the-Hole Hammer *Byull. Izobret.* No 31.

[5] Esin NN 1965 *Procedure for Investigation and Adjustment of Air Drill Hammers* Novosibirsk: SO AN SSSR (in Russian)

Causes of wear of PDC bits and ways of improving their wear resistance

VV Timonin[1], AS Smolentsev[1], I O Shakhtorin[1], NI Polushin[2], AI Laptev[2] and AS Kushkhabiev[2]

[1]Chinakal Institute of Mining, Siberian Branch, Russian Academy of Sciences, Novosibirsk, Russia
[2]National University of Science and Technology—MISIS, Moscow, Russia

E-mail: [1]timonin@misd.ru

Abstract. The scope of the paper encompasses basic factors that influence PDC bit efficiency. Feasible ways of eliminating the negatives are illustrated. The wash fluid flow in a standard bit is modeled, the resultant pattern of the bit washing is analyzed, and the recommendations are made on modification of the PDC bit design.

A rational approach to selection of a rock drilling tool ensures considerable savings of capital costs in drilling and construction of oil and gas wells, while optimization of drilling techniques promotes introduction of new types and designs of drill bits. In recent years, PDC bits increasingly gain popularity as they step up penetration rate per drill bit and cut down the trip in/out time, superseding roller cutting bits thereby [1].

When drilling in variable hardness rocks, drill bits experience with axial, radial and tangential vibrations. Under such conditions, drill bits continuously operate in percussion mode. The increment loads exerted on a drill bit increase the bit wear. Due to progressive wear of drilling tools, load per unit area of rocks lowers, which results in a decrease in the rate of penetration [2].

The key negative factors to affect drilling capability with PDC bits are wear of teeth, teeth holders and jet nozzles, reduction in bit diameter, defect of thread connections, vibration, packing and improper pattern of drill bit positioning.

Let us discuss each individual factor, starting with the most substantial influence and showing possible ways of eliminating negative impacts on drilling.

1. Wear of teeth. Among the listed above factors, this is the dominant factor for:

—first, wear of teeth governs coverage of teeth at well bottom, which conditions efficient drilling;

—second, wear of peripheral teeth conditions rated diameter of a drill bit;

—third, replacement of damaged teeth is the most expensive part of repair;

—fourth, evaluation of drilling characteristics of a drill bit consumes much time as it is necessary to examine individual teeth and, then, their ensemble, and to count the damaged teeth.

To assume a drill bit unworkable, it is required that more than 60% of teeth are damaged. When many teeth are worn-out, drilling performance drops while rate of wear of other members of a drill bit grows until complete failure.

From practice of drilling with drill bits, sides of a drill bit are subjected to major wear (see Figure 1a—wear of a drill bit with a diameter of 220.7 mm). Teeth arranged at different spacing along the radius from the drill bit axis deteriorate utterly unevenly. Peripheral teeth are in the most unfavorable conditions.

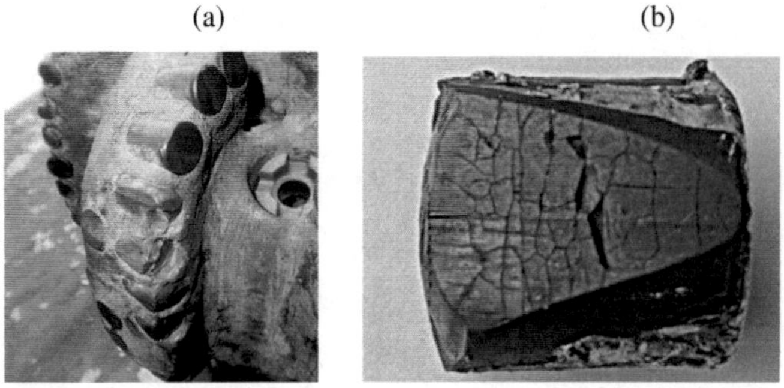

Figure 1. Wear of drill bits: (a) side of a drill bit; (b) fatigue grooves due to cyclical heating and cooling.

Wear of peripheral teeth is most often represented by chippage, galling, wear of underlayer with the concomitant symptom of thermal overload. Wear of drilling teeth after cutting in variable hardness rocks is of the mixed abrasive and impact-abrasive nature. Cyclical heating and cooling of surface during operation and, consequently, alternate expansion and contraction of surface layers causes the so-called fatigue grooves (Figure 1b).

It is known that downhole fluid motors are also sources of vibration due to disbalance of rotating rotor, clearings of radial supports and distortion of rotor shaft and housing. Amplitude of transverse vibrations may reach 10 mm and above, and frequency is equal to one vibration per revolution. High-rate wear of peripheral teeth is an aftereffect of this factor, too.

2. Wear of teeth holders. Replaceability of a tooth depends not on the depreciation of the tooth but on the integrity of the tooth holder characterized by percentage of damaged surface of the holder. Accordingly, when the surface of a holder is completely covered by the tooth body, housing and alloying material, the holder is assumed unworn (Figure 2a) [3].

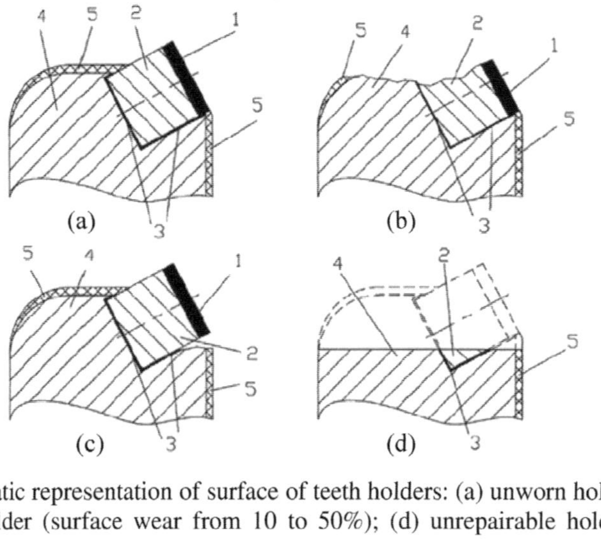

Figure 2. Schematic representation of surface of teeth holders: (a) unworn holder; (b) and (c) repairable holder (surface wear from 10 to 50%); (d) unrepairable holder (surface wear more than 50%); 1—diamond plate; 2—tangsten–carbide underlayer; 3—surface of tooth holder; 4—PDC bit body; 5—hard-alloy welding.

If the damage of a tooth holder makes 1/10 (i.e. visible) to 1/2 of the overall surface area, the holder is assumed repairable as it is yet possible to fix a new tooth in it.

Figures 2b and 2c illustrate typical partial damage of surface of a tooth holder with the lost portion of the tooth body or failed alloying material that protects the holder. Exposure and damage of surface of a holder usually take place simultaneously.

When more than a half of the holder surface is damaged (see Figure 2d), the holder is assumed unrepairable as it fails to preserve initial guiding base to orient a tooth.

3. Wear of jet nozzles. During operation of a dill bit, it is necessary to control jet nozzles and the absence of erosion. Under erosion or when any jet nozzle is lost, pressure difference in the bit reduces, washing fluid flow rate slows down and quality of cleaning of the dill bit and the well bottom deteriorates. Erosion of bit housing is understood as generation of holes in the housing, aside from drilling-bit fluid discharge ports.

The loss of a jet nozzle is sufficient to assume the jet nozzle unit is in want of repair. When a jet nozzle is come down, or initial diameter of the jet nozzle hole is eroded by 4 mm and more, the jet nozzle unit is unrepairable.

4. Reduction in bit diameter. This factor is an aftereffect of wear of peripheral teeth, which results in the gage loss of a drill well, and is to be controlled using a set of two specially manufactured gage rings with the diameters smaller than the rated diameter of a drill bit. Diameters of the first and second gage rings are selected based on geological features of the drill log, rigidity of the bottom of the drill-stem assembly and process steps scheduled for implementation in the course of drilling.

If a drill bit does not get through gage ring 1, the drill bit is assumed unreduced in diameter.

Passage through gage ring 1 and nonpassage through gage ring 2 defines the range of repairable drill bits. In this case, the operation of a drill bit is stopped, wear of other elements of the drill bit is evaluated and, based on this, a decision on the drill bit repair or disposal is made.

If a drill bit passes through gage ring 2, the drill bit gage loss is assumed and the drill bit is unrepairable.

5. Wear of thread connection. Any damage of thread, preventing from screwing a drill bit, is a sufficient condition for assuming the thread connection inappropriate for further operation of the drill bit.

6. Vibration. Vibration complicates the control over technical data of drilling, optimal drilling mode is failed and power consumed to rotate the drilling assembly sharply grows [2].

Among possible approaches to combating vibration, the most promising way seems to be installation of vibration dampers at the bottom of the drill stem assembly. It is highly important to stabilize a drill assembly during operation of drill bits, as this governs the drilling efficiency in variable hardness rocks and stabilization can ensure even loading of all teeth. Currently, bottomhole bumpers enjoy no wide application, and no studies have been carried out to examine PDC bit operation under conditions of vibration damping. In the East Siberia, bottomhole vibration dampers are the mandatory requirement for operation of drill bits. This method of damping vibrations is a promising technique to be used in drilling and construction of oil and gas wells in other regions of Russia.

Also, it is possible to suppress vibration of a drill stem assembly and to enhance efficiency of drilling in variable hardness rocks though installation of a hydraulic pulsator in the drill bit body.

The hydraulic pulsator placed in a counterbore in the drill bit body periodically catches drilling fluid in the drill bit cavity, pressure in the drill bit cavity grows, and opening of the cavity initiates an impulse in the drilling fluid flow. This allows transmission of impulses of axial dynamic loads directly to the drill bit with avoiding absorbing effect of intermediate units, including a downhole drilling motor. When a damper is installed between a drill bit and a downhole drilling motor, the impact exerted by the axial dynamic loads on the drilling motor is considerably reduced, which extends the motor service life. At the same time, transmission of high axial dynamic forces to a drill bit at the seam rpm enhances the cutting force of the drill bit, especially the vertical force component. Thus and so, with the increased cutting force, the rate of penetration grows.

7. Packing. This obstacle appears in PDC drilling in argillaceous deposits. As a consequence of adhesion of clay to metal surface of a drill bit, packing can coat entire drill bit, which totally blocks drilling (refer to Figure 3).

Packing is mostly prevented by modification of drilling mode (reduction in axial load on drill bit, increase in flow rate) or design of drilling tool. Also, it is possible to use drilling fluids with the improved antiadhesion characteristics.

In case of sticking of a drill bit, a borer has to try to remove a ball from a drill bit in the hole, or to move up an untreated drill bit.

There are some case histories when penetration stopped right after a drill bit was run down a well. It is thought that the cause of this is PDC bit packing as a result of its contact with the well walls. The chief problem in this case is chocking of space between blades with compacted rocks.

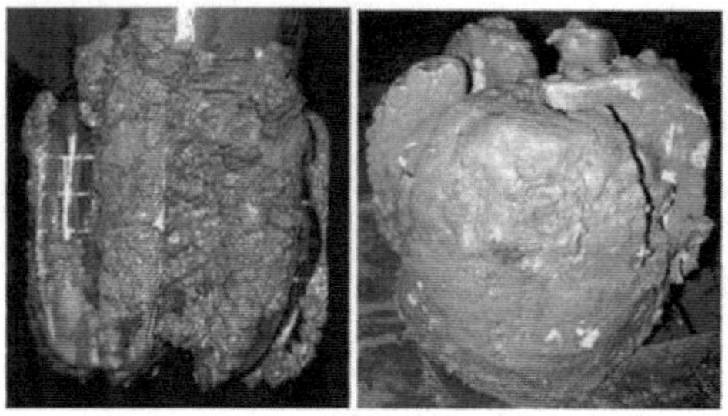

Figure 3. General view of a PDC bit with clay packing.

In this case, a drill bit is elevated above the well bottom by 15–20 cm, and pumps are switched on/off a few times. It is thought that fluid repelled from the bottom breaks down the packing by means of hydraulic impact; however, this method is not always effective in cleaning a drill bit [4].

8. Pattern of drill bit positioning. There single-station, many-station and multi-position drill bits.

In case of a single station PDC drill bit, each tooth of the bit is given unique radial position determined relative to the central bit axis outward in the direction of the bit diameter. One of the common methods of single-station drill bit arrangement is placement of teeth at the intersections of the spiral line, drawn from the drill bit axis, and each blade. In this case, each intersection point of the spiral and blades occurs at different radial distance from the drill bit axis.

Single-station PDC drill bits demonstrate higher rate of penetration; however, when a tooth is damaged or lost, wear of the nearest radial teeth quickens. This results in early failure of a drill bit.

In many-station drill bits ("backup tooth" or "accompanying tooth" schemes), teeth are actuated in groups of two or more teeth, and teeth in a group are placed at the same radial distance from the drill bit axis. In view of smaller area, fewer blades can be arranged at the drill bit end, and not every tooth can be a member of a group arranged along the same radius; and still, most of teeth belong in the same group. A typical pattern of teeth is when grouped teeth (along the same radius) are arranged, e.g., on opposite blades.

Multi-position PDC drill bits are, as a rule, highly wear resistant as against single-station bits but exhibit slower penetration rates [5].

A multi-position drill bit is better balanced as compared with the single- and many-station bits. With the closely spaced backup secondary blades in combination with the equivalent arrangement of backup teeth along the contour of a drill bit, the backup teeth cut minimum amount of rock per each rotation of the drill bit. Single-station primary teeth on the primary blades fulfill much work, and the backup teeth on the secondary blades only cut groove bottom after the primary blade.

All factors discussed above exercise considerable influence on efficiency and service life of PDC drill bits. Furthermore, reduction in wear of members of a drill bit is feasible through correct operation of mud system, which is intended to ensure both timely removal of chipping and prompt and efficient cooling of cutting tools.

It is known that drilling temperature at the cutting tool and rock contact may reach 1000°C, and inefficient cooling can cause critical wear of diamond-bearing cutting layer and underlayer, and can induce fall of teeth from holders.

Efficient cooling of cutting tools requires that a mud system ensures required and sufficient washing fluid flow rate and high flow rate along active faces of blades.

Aimed at adjustment of drill bit design and determination of rational geometry with a view to enhancing drilling efficiency, the process of flushing-out of well bottom has been modeled using SolidWorks Flow Simulation [6].

After the analysis of the model of washing fluid flow in a drill bit (Figure 4), it has been concluded that:

1. The mud system does not ensure direction flow of washing fluid (Figure 4, positions 1 and 4), which results in poor cleaning of well bottom, in probability of extra interaction with chipping and, eventually, increased wear of cutting tools.

2. The inter-blade space has an expanding cross-section, which conduces to speed drop Figure 4, position 4) and turbulence (Figure 4, position 1) of washing fluid in the inter-blade space, and causes packing.

3. The washing fluid flow rate is high along the well walls (Figure 4, position 5) but is lower along the active faces of blades (Figure 4, positions 2 and 6), which ends with insufficient cooling of cutting tools.

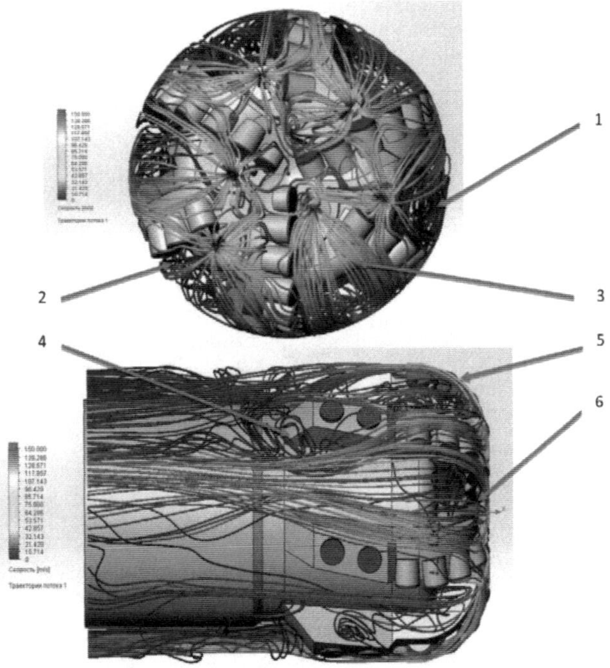

Figure 4. Velocity of washing fluid at the flow rate of 30 l/min.

4. The main flow of washing fluid (Figure 4, position 3) is directed to the well bottom, along the inactive face of the blade, which is worsens heat withdrawal.

Elimination of the listed shortcomings requires to increase washing fluid velocity along the active faces of blades of PDC drill bits. Efficiency of cooing of cutting elements will grow in this case, and wear of cutting tools will abate as a result, which will extend service life of drill bits. It is possible to speed-up washing fluid flow at the periphery of the inter-blade space by ensuring uniform cross-section of the inter-blade space to prevent packing and to improve removal of chippings.

References

[1] Chulkova VV 2012 Experience of drilling with PDC drill bits in variable hardness rocks in the Ural–Volga region *Bulletin of Association of Drilling Contractors* (in Russian)
[2] Chulkova VV 2012 Resource-saving drilling with PDC drill bits under conditions of variable hardness rock mass *Burenie Neft*, pp 56–57.
[3] Myasnikov YaV, Ionenko AV, Gadzhiev SG et al. 2014 Guide on PDC drill bit wear evaluation in field conditions *Burenie Neft* No 3, pp 14–18.
[4] http://www.buroviki.com/yaf/forum.aspx?g=posts&m=584
[5] *Multi-Set PDC Drill Bit and Method* Inventors: Drews SW, Maurstad C. Assignee: Varel International, Ind., L.P. 2012
[6] Timonin VV and Kharlamov YuP 2014 Mathematical modeling of mud-circulating system of an air drill hammer with a central mud removal line using SolidWorks Flow Simulation, *Vestn. KuzGTU*, 6 (106), pp. 61–65.

Simulation modeling and tracing optimal trajectory of robotic mining machine effector

VN Fryanov* **and LD Pavlova****

Siberian State Industrial University, Novosibirsk, Russia

E-mail: *zzz338@rdtc.ru, **lara@rdtc.ru

Abstract. Within the framework of the robotic coal mine design for deep-level coal beds with the high gas content in the seismically active areas in the southern Kuzbass, the motion path parameters for an effector of a robotic mining machine are evaluated. The simulation model is meant for selection of minimum energy-based optimum trajectory for the robot effector, calculation of stresses and strains in a coal bed in a variable perimeter shortwall in the course of coal extraction, determination of coordinates of a coal bed edge area with the maximum disintegration of coal, and for choice of direction of the robot effector to get in contact with the mentioned area and to break coal at the minimum energy input. It is suggested to use the model in the engineering of the robot intelligence.

Reaching of economic efficiency and environmental safety of underground coal mining will condition the role of coal in the energy strategy of Russia in 5–10 years to come. The relevance of this objective is conditioned by the competition on the market of alternative energy sources, fall in oil and gas prices and anticipated imposition of a "carbon" tax for CO_2 emissions.

The reasons for abridging underground coal mining include [1]:

—introduction of alternative ecology-friendly sources of energy;

—decision of the Climate Change Summit in Paris in 2015 on the mandatory reduction in toxic gases, including in Russia, in the years immediately ahead by 30% as against emissions in 1990 with a view to decelerating the rate of warming of the Earth's climate;

—complication of natural conditions of underground mining: greater depths down to 900 m below ground surface, higher gas content of coal beds over 13 m^3/t, elevated risk of gas-dynamic events;

—high accident rate in Russian mines: 11 large accidents for the latest 12 years, with more than 350 workers died.

For the first turn, as in the coal industry restructuring in 1992–2000 in Russia, it is planned to close coal mines with the super high methane content, prone to spontaneous firing and gas-dynamic events.

One of the ways of enhancing efficiency and safety of underground geotechnology in complicated natural environment is replacement of mine personnel with robotic machines.

With this end in view, the Siberian State Industrial University has developed the scientific basis for the robotics technology for deeper level mining of coal beds having high gas content [1–3].

The essentials of the proposed robotics geotechnology are:

—traditional methods of getting access to coal and mine field preparation;

—room-and-pillar method, with ventilation, parallel and accumulating drifts driven around a panel (see Figure 1);

—hydraulic mining of coal broken by high pressure water jets, hydraulic transport of coal slurry to underground dewatering chamber and traditional elevation of coal to ground surface.

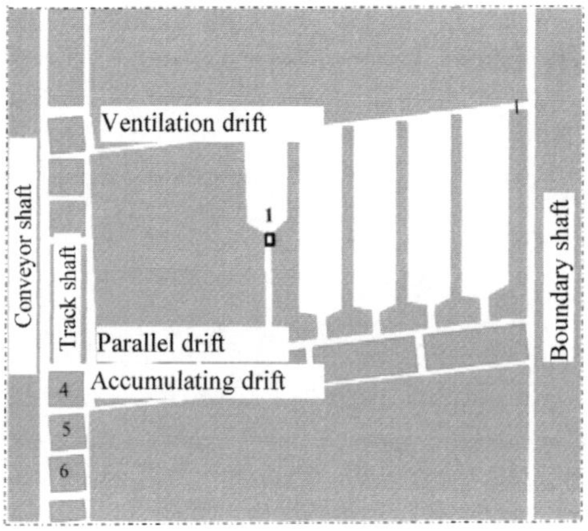

Figure 1. Automated coal mining with short rooms.

The designed robot-machine system is capable of:

—forward driving with hydraulic breakage of coal and gravity handling of coal into accumulating drift;

—backward widening of the stall, with turn of a jet generating box with nozzles in perpendicular to the room axis and with gravity handling of coal into accumulating drift;

—support of roof rocks at the stall and face junction with a remote canopy;

—remote and computer-aided control of the processes and operations from operator's panel.

Hydraulic breakage of coal eliminates such processes and operations involved in the conventional geotechnology as:

—mine ventilation as coal extraction is implemented without mine personnel present in the face area;

—precaution of inflammation and explosions of methane in the production face area, as mine air is phlegmatized by water vapors;

—hand control of mining machines due to remote monitoring of state of the production face area and owing to computer-aided control of the robot-machine system from operator's panel located in safe area, including ground surface;

—conveyors and locomotives within the limits of an extraction area owing to gravity hydraulic handling of coal slurry to a dewatering chamber;

—roof rock support as stability of sidewalls is ensured by minimized round-shaped cross section.

As is known, hydraulic coal mining technology used in the 1960s–90s features some drawbacks [4–6], including:

—accomplishment of basic and auxiliary operations by hand;

—large loss of coal, making 15.4–34.3% in 1988;

—high energy input for mining, hydraulic transport and hydraulic elevation of coal: energy consumption was 410–655 MJ/t of hydraulic coal mines using high-pressure jets and 182–464 MJ/t in water-jet-assisted mining in 1988.

The new proposed robotic mining technology with coal breaking by thin high-pressure jets will allow partly eliminating the above listed drawbacks. This paper presents some data on testing of coal breakage by fine jets with minimized energy input of mining due to optimized trajectory of the effector of the mining robot-machine system pre-set based on points of the maximum energy of rock pressure. The assumed optimization criterion for the coordinates of a current breakage area is given by

$$W_G + W_H \geq W_T, \quad (1)$$

where W_G is the specific energy of coal breakage under integrated influence of natural and mining-induced stresses generated by geological and geotechnical factors; W_H is the specific energy of coal breakage under influence of high-pressure hydraulic jets; W_T is the theoretical specific energy of coal breakage under laboratory conditions, with regard to structural heterogeneity and boundary conditions of a coal specimen to be in conformity with mine conditions.

To find the specific energy of coal breakage under high-pressure hydraulic jets, W_H, the data of the laboratory tests from [7] were used, and the relation below was obtained after transformation of coefficients of dimension of values

$$W_H = A\left(\frac{f}{P_j}\right)^2, \quad (2)$$

where f is the hardness of coal by Protodyakonov's scale; P_j is the jet pressure on the surface of the edge area of coal bed under breakage, MPa; A is the index of jointing of coal: recommended value is 20 for heavily jointed coal, 40 for medium-jointed coal and 60 for weakly jointed coal.

The theoretical value of the specific energy for coal destruction, W_T, was determined with due regard to the uniaxial strength of coal in the edge area from the formula [8]

$$W_T = \frac{(1+v)v\sigma_c^2}{E}, \quad (3)$$

where v is Poisson's ratio of coal; E is Young's modulus of coal; V is the coal volume under breakage, m^2; σ_c is the compressive strength of coal.

The breakage energy W_G under geological and geotechnical factors was found from numerical modeling of stress state of a coal bed, considering shape and dimension of a panel and a stall.

The modeling used a dedicated application package designed by the authors [9] and ensuring 3D problem solution in rock mechanics. The input data of the geometrical model of coal and rock mass were: depth of mining—900 m, total coal bed thickness—3 m, coal bed dip—12 deg; ultimate compressive strength of coal—10 MPa, extraction panel—200 m along the dip and strike, room was round-shaped with a diameter of 1 m and 6 m wide; rib pillar width was 1 m.

Rocks of immediate roof 14 m thick were represented by alternating argillite and siltstone with the ultimate compression of 20–40 MP; main roof with the thickness of 30 mm was composed of siltstone with the ultimate compression of 40–60 MPa. The shortwall parameters were assumed in conformity with the research results from [10].

The modeling included relocation of the face in the course of expansion of the shortwall relative to the west boundary of the extraction panel with the elongation of the mined-out void and the stall (see Figure 1).

Based on the modeling data, total vectors of stresses, strains and displacements in rocks, ratio of residual strength to natural strength of coal and deformation energy were calculated.

Figures 2 and 3 show the isolines of vertical stresses and specific energy of coal deformation at the juncture of the shortwall and stall. The similar plots were obtained for various locations of the face within the extraction panel.

According to Figure 2, the vertical stresses are extremely unevenly distributed in the immediate roof and, consequently, in the coal bed. The line connecting the points of the maximum abutment pressure goes along the diagonal relative to the stall axis. In conformity with the hypothesis of elastic deformation, a rock burst is expected in the vicinity of the stall. This proves the necessity of remote extraction of coal in shortwalls.

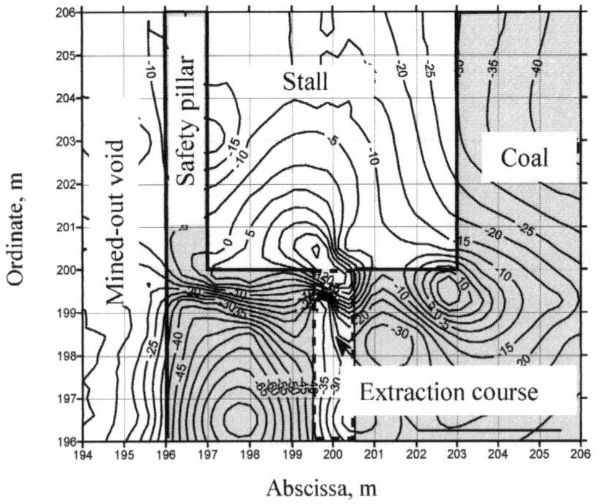

Figure 2. Isolines of vertical stresses (MPa) in the immediate roof rocks at the juncture of the shortwall and stall.

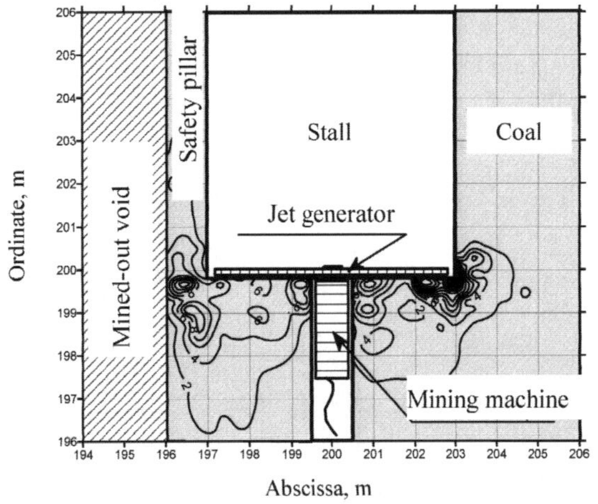

Figure 3. Isolines of the deformation energy (MJ) in coal bed at the juncture of the shortwall and stall.

At the shortwall and stall juncture, in the immediate roof rocks, tension stresses can induce rock falls and roof caving. This is also a fact in favor of the robotic coal extraction.

The deformation energy is also extremely unevenly distributed (Figure 3), which requires coal extraction with the minimized energy input with the prediction of the coordinates of the maximum energy deformation points in the production face area.

The modeling allowed finding that stresses in coal–rock mass and the deformation energy in edge are of coal greatly depended on:

—mining depth and coal bed thickness;

—shape and dimension of mined-out area;

—location of the production face area relative to the boundaries of the extraction panel;

—width of the coal pillar between neighbor stalls;

—coordinates of hydraulic mining zone on the surface of the coal bed.

The flow scheme of the coal mining control system is shown in Figure 4. The system operation optimization criterion is the minimum breakage energy W_H under the action of high-pressure jets.

In accordance with (1), the objective function of the specified optimization criterion is

$$W_H = W_G - W_T \to 0. \quad (3)$$

The system of the remote control of coal shortwall mining includes:

Block 1: Object of control, which is a shortwall equipped with a robotic machine system as is shown in Figure 2.

Block 2: Subsystem of continuous remote monitoring of the shortwalling parameters (breakage efficiency, hydraulic transport of water and coal slurry, varied coordinates of the production face surface and jet nozzles in the course of mining, roof rock deformation, mine air parameters, etc.);

Block 3: Bundled software to estimate current conditions and predict optimal trajectory of the mining robotic machine effector, considering the objective function (4), including:

—calculation of stress state of coal bed with regard to the change in the production face coordinates;

—selection of the production face coordinates towards the maximum energy of coal breakage under integrated natural and mining-induced stresses;

—calculation of the coordinates of nozzles of high-pressure jets in accordance with (3) and (4);

—calculation of high-pressure jet parameters (pressure, number of nozzles, water flow rate).

Block 4: Subsystem of decision-making by an operator or a robot in accordance with the pre-set program.

Block 5: Subsystem of remote transition of control to a robotic mining machine.

The remote control flow chart for coal shortwalling is proposed to be used in engineering a mining robotic mining machine.

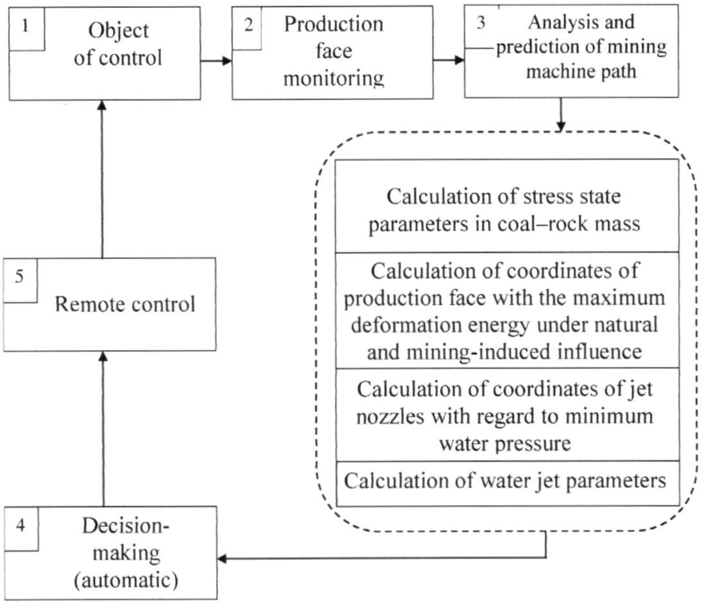

Figure 4. Flow chart of remote control of coal shortwalling.

In actual coal mining control, calculation of stresses and energy W_G of coal deformation under influence of geological and geotechnical factors in each cycle of change in the position of the production face and decision-making as per Block 4 (Figure 4) will consume much time. Aiming to enhance operating efficiency of the production face control, it is proposed to carry out simulation modeling at stage of development of an extraction panel specifications and standards and to generate a matrix in the form of 3D digital model of geomechanical parameters of coal–rock mass. A fragment of such matrix is given in the table below with regard to the production face coordinates in Figure 3. The negative values of the deformation energy in the table correspond to the coal disintegration conditions under energy of rock pressure, i.e. no high-pressure jet treatment is required in this case, and only low-pressure water is to be fed to the production area to ensure gravity hydraulic handling of broken coal and rocks.

Fragment of a digital model of geomechanichal and geotechnical parameters of a production area in coal bed with the ultimate stress limit of 10 MPa.

Breakage zone coordinate in the face area, m			Deformation energy, MJ		
X	*Y*	*Z*	W_T	W_G	W_H
1	2	3	4	5	6
Distance to the face 1.5–2.5 m, ultimate compression strength of coal is 20 MPa					
196.25	197.75	–2	6.68	1.39	5.29
196.75	196.5	–2	6.68	2.02	4.66
197.37	195.75	–1.75	6.68	1.52	5.16
198.00	196.5	–1.8	6.68	1.64	5.04
199.00	197.25	–0.88	6.68	1.38	5.3
201.25	198.5	–0.5	6.68	5.16	1.52
202.25	198.5	–1.5	6.68	4.3	2.38
203.25	198.25	–1.3	6.68	1.71	4.97
Distance to the face 0.6–0.9 m, ultimate compression strength of coal is 15 MPa					
196.25	199.12	–2	3.75	5.38	–1.63
196.75	199.25	–2	3.75	8.29	–4.54
197.37	199.25	–2	3.75	4.28	–0.53
198.00	199.37	–0.88	3.75	3.72	0.03
1	2	3	4	5	6
199.00	199.37	–1.75	3.75	6.94	–3.19
201.25	199.37	–1.8	3.75	5.5	–1.75
202.25	199.37	–1.8	3.75	7.46	–3.71
203.25	199.37	–1.8	3.75	2.42	1.33
Distance to the face 0.1—0.5 m, ultimate compression strength of coal is 10 MPa					
196.25	199.63	–2	1.67	7.55	–5.88
196.75	199.87	–2	1.67	3.85	–2.18
197.37	199.87	–2	1.67	3.35	–1.68
198.00	199.87	–0.88	1.67	6.68	–5.01
199.00	199.87	–1.75	1.67	4.67	–3.00
201.25	199.87	–1.8	1.67	10.46	–8.79
202.25	199.87	–1.8	1.67	20.04	–18.37
203.25	199.87	–1.8	1.67	9.34	–7.67

Using the numerical modeling results, the curves of the deformation energy in coal under influence of rock pressure at various distances to a production face are plotted in Figure 5. The dotted and dashed lines show the limit energy required for coal breakage. At the points located above the limit energy values, it is not required to use high-pressure jets.

It is suggested to display such plots on an operator's panel for making control decisions (Figure 4).

It is recommended to use the research findings in development of technical requirements for layout and software support of a robotic mining machine.

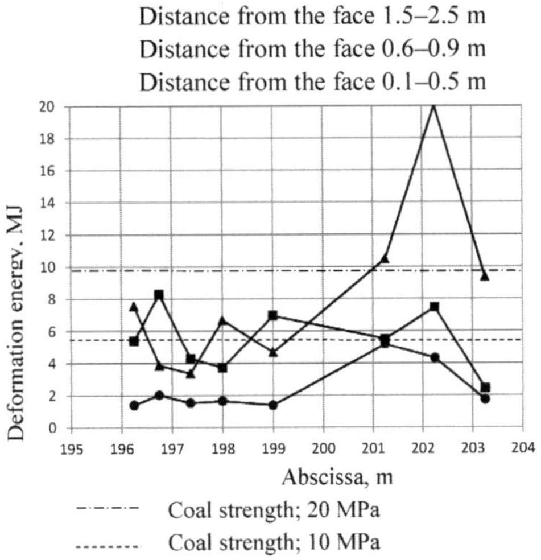

Figure 5. Energy deformation of coal under rock pressure.

Conclusion

The relevance of robotic coal extraction technology to be developed in introduced in mining in complicated natural conditions has been substantiated.

The authors have put forward a process flow chart of coal mining with a robotic machine in gassy area using high-pressure jets and hydraulic transport of broken coal to a dewatering underground chamber.

The procedure to select optimal parameters for high-pressure jet operation is based on the criterion of minimized destruction energy at maximum utilization of energy of rock pressure.

References

[1] Fryanov VN, Pavlova LD 2016 Prospects for robotic hydraulic mining of deep-level coal beds with high gas content *Science-Driven Technologies for Mineral Mining and Use: Int. Conf. Series* No. 2 (in Russian)

[2] Fryanov VN, Pavlova LD 2013 Geomechanical and gas-dynamic information support of a robotic coal mining technology *Science-Driven Technologies for Mineral Mining and Use: Int. Conf. Series* pp. 5–11 (in Russian)

[3] Pavlova LD, Fryanov VN, Kornev ES 2011 Prediction of geomechanical parameters in robotic coal mining *Geodynamics and Stress State of the Earth's Interior: Int. Conf. Proc.* Novosibirsk: IGD SO RAN pp 163–169 (in Russian)

[4] Muchnik VS, Golland EB, Markus MN 1986 *Underground Hydraulic Coal Mining* Moscow: Nedra (in Russian)

[5] Atrushkevich AA, Kaido I.I., Fomichev SG 1999 Hydrotechnology—Economically efficient coal mining *Ugol* No. 10 pp. 63–64

[6] *Report on Performance of Hydraulic Mines and Dewatering and Preparation Plants in Kuzbass and Donbass in 1988* Novokuznetsk: VNIIgidrougol (in Russian)

[7] Tsyapko NF 1957 Research findings on hydraulic coal mining *Proc. I All-Union Conf. on Hydraulic Coal Mining* Stalinsk: VNIIGidrougol (in Russian)

[8] Timoshenko S, Goodier JN 1951 *Theory of Elasticity* McGraw Hill Book Company

[9] Kornev ES, Pavlova LD, Fryanov VN 2013 Design of a system of problem-oriented programs for finite eleemnt modeling of geomechanical processes *Vestn. KuzGTU* No. 2 pp 65–69

[10] Fryanov VN, Egorov VP, Kovalev VA, Slavnikov VD 1999 *Geomechanical Control in Coal Mining with Shortwalls* Kemerovo: Akad. Gorn. Nauk (in Russian)

Borehole tool for studies in coalbed degasification wells

SV Serdyukov, AV Patutin[*] and TV Shilova

Chinakal Institute of Mining, Siberian Branch, Russian Academy of Sciences, Novosibirsk, Russia

E-mail: [*]andrey.patutin@gmail.com

Abstract. The paper presents a downhole tool designed for gas-dynamic research to be carried out in coalbed methane drainage holes. Structurally, the tool has a twin-packer design. The tool allows hydraulic fracturing, gas-dynamic investigations using indicator diagrams and pressure drop and recovery curves, and local destressing of coal.

A goal of gas-dynamic studies of degassing boreholes is to determine coalbed gas-saturation and permeability, to find vigorous gas evolution zones which make mining complicated. A problem of the studies is mismatch in conditions of gas escape into a borehole or an excavation. In case of interval borehole survey, exposed surface of coal bed is small. Due to stress concentration at the borehole boundaries, coal undergoes higher pressure than enclosing rock mass. Near-surface bed is damaged and partially degassed during the process of drilling. In order to increase the accuracy of the studies we suggest conducting them in coalbed hydraulic fracturing interval along with the radially symmetric loading of the borehole walls. The loading provides hydraulic fracturing crack opening, so it is possible to measure gas recovery of the coalbed, when degassing area surface free from stress is enlarged. After a degassing borehole interval is sealed, the crack opens, which provides the opportunity to study gas recovery from the beginning of degassing surface formation out of drilling influence zone.

This article describes methodological and technical solutions providing the interval gas-dynamic studies along with the regulated coalbed destressing. On the basis of the solutions we have developed a borehole tool equipped with embedded system of travel along the borehole and flexible high-pressure hoses connecting the borehole equipment with the mine working equipment (pump, remote control, measuring system). The tool is capable of performing the following technological operations and measurements:

—sealing of an uncased borehole interval, supplement of a working agent (compressed nitrogen, water-in-oil emulsion) under pressure or drawing off formation fluids controlling their pressure, temperature and volume discharge;

—hydraulic fracturing of coalbed by the working agent and measuring of lock-up pressure P_s and reopening P_r of the created crack;

—radially symmetric loading of hydraulic fracturing interval through plaster cover preventing the contact of working fluid and rock mass;

—increase or decrease of pressure in the isolated interval of the borehole and the following measurement of decline curve or pressure P(t) recovery curve;

—determination of indicator diagram (ID)—dependence of pressure in the borehole isolated interval on formation fluid (gas, water) volume charge Q.

The tool scheme is shown in Figure 1.

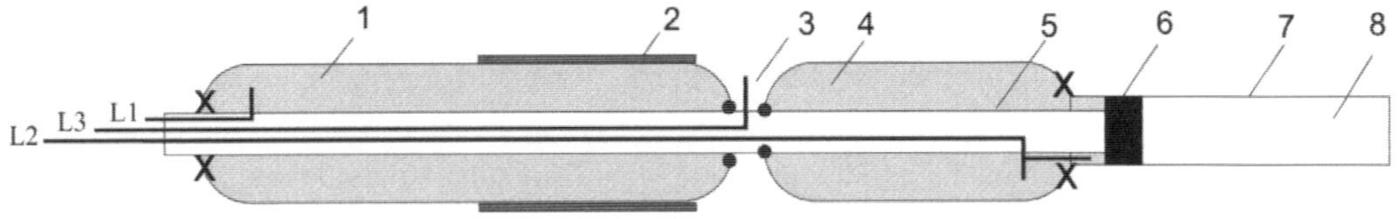

Figure 1. Borehole tool for gas-dynamic studies: *1*—packer sleeve, fixed at hydraulic-cylinder rod; *2*—drainage layer; *3*—injection zone; *4*—packer sleeve, fixed at hydraulic-cylinder body; *5*—hydraulic-cylinder rod; *6*—hydraulic-cylinder piston; *7*—hydraulic cylinder; *8*—gas spring; X—fixed sleeve end; • – movable end; L1–L3—high-pressure hoses.

The tool is made accordingly to the scheme of a double packer with controllable interval *3* between cord-reinforced air bags *1* and *4* (Figure 1). Regulation of injection zone *3* is performed with the help of hydraulic cylinder *7*. There is an immovable sleeve end *4* fixed to its body. Long air bag *1* is fixed on polished rod *5* of the hydraulic cylinder. Sealed movable ends of both packer sleeves shift along the rod. The main technical characteristics of the tool:

Tool length	2870 mm
Tool diameter	60 mm
Tool mass	12 kg
Hydraulic fracturing pressure	Up to 18 MPa
Borehole wall loading pressure by packer	Up to 26 MPA
Borehole diameter	76–105 mm
Borehole length	Up to 1000 m
Borehole orientation	No limit
Continuous work duration	No limit
Travel speed in borehole	Up to 100 m/h

The tool works in three modes: transition in the borehole, hydraulic fracturing, crack opening by the packer sleeve and gas-dynamic study. In the transition mode the tool works according to the following scheme. The working agent under pressure is supplied to the packer sleeve *4* and hydraulic cylinder *7* through the high-pressure hose L2. Under the influence of the working agent pressure the packer sleeve *4* is inflated and adherent to the rock mass. In this state the packer sleeve works as an anchor for the cylinder body *7*. The working agent presses on the piston *6* and moves it with the hydraulic cylinder rod *5* into the borehole, pulling up the packer sleeve *1* and high-pressure hoses L1–L3. Then, through the high-pressure hose L1 the working agent is supplied into the packer sleeve *1*, which inflates and, therefore, contacts the borehole walls. Through the line L2 the packer sleeve *2* pressure is transferred into the drain system. The sleeve *2* does not have the contact with the ground. The hydraulic cylinder gas spring *8* influences the sleeve *2* so that relatively to the fixed rod the sleeve moves into the borehole.

In case of hydraulic fracturing the tool works in another way. First, the working agent is supplied into the packer sleeve *2*, and then, into two sleeves *1*, 2 under the pressure of 2–4 MPa. This order of operations provides the opportunity to gain the maximum injection distance. Next, the working agent is supplied into two sleeves, and through the pressure reducing valve and high-pressure hose L3 the working agent is pumped into hydraulic fracturing interval, until a crack is formed in the rock (Figure 2a). The valve helps to maintain the pressure ratio 4–6 MPa between the packers and the injection interval. These conditions are sufficient to keep the fractured interval sealed. A sensor measures pressure in high-pressure hose L3. On the basis of its measurements we determine P_s and P_r pressures, which are used to estimate the rock mass unloading and area of its degassing.

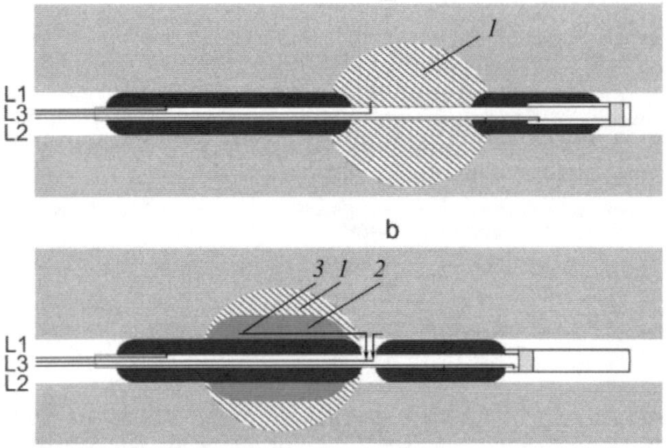

Figure 2. Scheme of the tool working in different modes: hydraulic fracturing (a) и gas-dynamic research (b): *1*—hydraulic fracturing crack; *2*—area of the crack opening by the packer sleeve; *3*—gas delivery into the injection interval.

In the gas-dynamic research the tool works in the following way. At first the tool is placed into the borehole so that a part of the long packer sleeve *1*, coated with drainage layer, takes place in the interval of the hydraulic fracturing crack (Figure 2b). After that, under 2–4 MPa pressure the working agent is supplied into the packer sleeve *1*, then—into two sleeves *1*, 2. Thus, by preserving this order we get the minimal injection interval length.

The tool provides the opportunity to conduct gas-dynamic research of the coalbed zone in steady (indicator diagrams ID) and non-steady (decline and recovery pressure curves) states. Implementation of different methods in the tool increases the measuring accuracy of permeability and storage properties of a coalbed. Gas-dynamic research conducted by any of the mentioned methods is carried out if there is a hydraulic fracturing crack opening. To cause the opening we set pressure bigger than P_r in packer envelope *1*. The depth of the crack opening is determined by the methods presented in the work [*1*].

Data processing is performed in the following way:

—in accordance with the measurements plot the dependence $(P_0 - P)/Q_g$ on gas output flow Q_g, where *P*—pressure in injection interval; P_0—formation pressure defined by pressure stabilization in closed injection interval when $Q_g = 0$;

—approximate the curve by dependence $\frac{P_0^2 - P^2}{Q_g} = a + bQ_g$ and determine *a* and *b* coefficient using the ordinary least squares method;

—calculate the expected maximum gas rate when the borehole is set with degassing system of a mine by the formula $Q_g^{\max} = \frac{\sqrt{a^2 + 4b(P_0^2 - P_V^2)} - a}{2b}$, where P_V – pressure in the mine degassing system;

–calculate the averaged gas formation pressure $\overline{P}_0$ using the formula

$$\overline{P}_0 = \frac{1}{N}\sum_{n=1}^{N}\sqrt{P_n^2 + aQ_{g_n} + bQ_{g_n}^2}\,. \qquad (1)$$

If the obtained formation pressure differs from the value used for the indicator curve construction, then, the averaged pressure calculated from the formula (1) is used for the second data processing accordingly with the mentioned algorithm.

The measurements of the decline pressure curve are processed by the following simplified technique:

—with the help of the conducted research determine the gas flow before the research and schedule a dependency graph of the square of the pressure in the injection interval on the common logarithm of time t, counted in seconds from the pressure recovery;

—approximate the graph by the dependence $P^2 = \gamma + \beta \lg t$, where γ and β are the approximation coefficients;

–using the obtained value β calculate the gas permeability of the coal bed using the formula $\varepsilon_g = \frac{0.023 q T_0 Z_0 P_V}{\pi \beta T_s}$, where q—gas flow rate before the research conducted by decline pressure curve, m^3/s; T_0 – gas temperature in the bed, K; $T_s = 293$ K; Z_0—methane compression coefficient in case of bed's temperature and pressure;

—using the well-known gas dynamic viscosity in beds μ_0 calculate product of horizontal permeability and net pay thickness on gas transferred to injection interval's length unit h

$$k_g = \frac{\varepsilon_g \mu_0}{h}.$$

Data can be processed with the other improved methods not discussed in the article. It should be pointed out that the coalbed can contain methane in bound and unbound states. When the bed is being drilled the unbound gas goes into the low-pressure zone and the bound gas diffuses in the filter volume of the coalbed. As a result there are two areas on the pressure recovery curve; each of them is studied separately. As the diffusion process is relatively slow, the test can take from several hours up to several days, which is not possible in case of a borehole study. The alternative is the assessment of the unbound methane gas recovery. In this case total gas content of the coal bed can be determined by the estimation of the unbound and bound methane relations in the bed, for example, by the methods mentioned in the paper [2].

Conclusion

We have developed a tool providing the opportunity to conduct the gas-dynamic research of a coalbed in steady and non-steady states using indicator diagrams, pressure decline curves and pressure recovery curves. The tool ability to combine different techniques increases the validity of the permeability and the storage capacity of a coalbed, including its formation pressure, gas permeability and expected rate of degassing boreholes.

A distinctive feature of the tool is its capability to conduct research at a distance from a borehole collar. For this purpose the tool is equipped with embedded travel system in uncased extended borehole of any orientation. The tool is capable of conducting research being several hundred meters away from the mine working.

The tool provides hydraulic fracturing of a coalbed by working agent and gas-dynamic research with the formation of a crack due to radially symmetric loading of borehole walls. Local loading of a coalbed increases the accuracy of the research into coal gas content and makes it possible to study gas recovery depending on degassing surface area free from pressure.

References

[1] Martinyuk PA, Pavlov VA and Serdyukov SV 2011 Assessment of stress state in rocks by deformation characteristic of borehole zone with hydrofracture *J. Min. Sci.* Vol 47 No 3 pp 290–296

[2] Feldman EP, Vasilenko TA and Kalugina NA 2014 physical kinetics of coal-methane system: mass transfer, pre-outburst events *J. Min. Sci.* Vol 50 No 3 pp 448–464

Unbalanced rock mass vibration generator and its bench testing

LA Rybalkin[1], AV Azarov[1] and AS Serdyukov[1,2]

[1]Chinakal Institute of Mining, Siberian Branch, Russian Academy of Sciences, Novosibirsk, Russia
[2]Institute of Petroleum Geology and Geophysics, Novosibirsk, Russia

E-mail: :[1]leonid.rybalkin@gmail.com, [2]serdyukovas@ipgg.sbras.ru

Abstract. This work presents an unbalanced borehole vibration source with the pneumatic actuator to affect the rock mass bottomhole zone in the seismic frequency range. The modulus-type source consists of the vibration generator with automatic step regulation of static moment, clamping node and built-in pneumatic device to transfer equipment in uncased boreholes. There are test results on the source prototype, amplitude and frequency characteristics and spectral content of the source signal.

1. Introduction

Seismic impact on rock mass is used to increase permeability and fluid flow in rocks by low-frequency elastic vibration treatment. The main object of the seismic impact for borehole oil and gas production technologies, coal bed degassing and nonferrous metal in-situ leaching is a bottomhole area of producing wells. The area hydrodynamic properties affect production capacity of the wells. One of the main factors of successful production is specific capacity of seismic emission dependent on the source type and its properties.

Experimental investigations prove that in order to increase gas and water permeability of a bottomhole area, impact frequency has to range from 100 to 300 Hz, intensity—not less than 0.01 W/m^2, duration—from 4 hours and more [1–4]. Numerical studies of near zone of different source emission [5–7] shows that the required values of the mentioned parameters are ensure by an unbalanced seismic source (unbalanced vibration generator) with the force 10–20 kN and higher [5]. Design of the source and results of its tests in soil and hard rocks are given below.

2. Unbalanced borehole source

The source contains unbalanced type generator of elastic oscillations *1* with a pneumatic actuator and electromechanical pressure device *2* (Figure 1). The source is capable of working along with the pneumatic device used for the transportation of equipment in a borehole [8]. Other facilities are remote control, air supply hoses, conducting wire, frequency counter and geophone set next to the borehole collar. The geophone is employed to control the source operation.

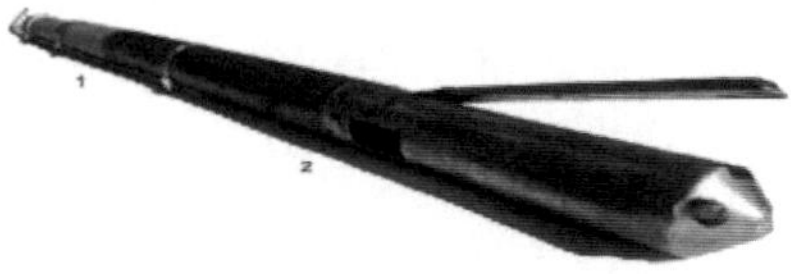

Figure 1. Borehole vibration generator.

In the transportation mode pressurized air is supplied into the pneumatic device which transfers the equipment into the depth of the borehole [8]. In case of seismic impact pressurized air enters the pneumatic actuator input, and it is discharged into the atmosphere through the output. Vibration frequency is regulated by the frequency counter data. Signals of geophone, registering rock displacements, are transmitted to the counter input. Frequency control is accomplished due to the regulation of pressure and air supply.

2.1. Pneumatic actuator

We have chosen a type and power of a pneumatic actuator of the unbalanced generator on the basis of seismic impact intensity equal to 0.01 W/m^2. If the radius of vibration impact is limited to 10 m, then, the seismic emission intensity has to be not less than 12 W. At the same time it is a well-known fact that in rock masses energy conversion coefficient of seismic transmitters does not exceed 5–10% [9]. To obtain the required seismic energy the actual power has to be not less than 120–240 W. In the limited space of a borehole this power is provided by volumetric pneumatic actuator made in accordance with high-technology pneumatic motors with 40–50% energy conversion efficiency. The unbalanced generator is embedded with PHR motor (Ober, Italy) with the following parameters: up to 570 W of output power, 316 r/sec of maximum rotation speed without loading, 183 r/sec of maximum rotation speed if power is 570 W, up to 13.8 l/sec of gas supply (input discharge) in normal conditions and 0.62 MPa pressure, 0.8 N·m of torque, 1.1 kg mass and 50 mm diameter.

The air hose provides the required pressure and air supply into a pneumatic motor. To calculate a diameter of air hose flow section *d* we use Weymouth formula for simple gas pipeline [10]

$$d = \left[\frac{a \cdot V^2 \cdot \rho_a^2 \cdot z \cdot R_a \cdot T \cdot L}{P_1^2 - P_2^2} \right]^{3/16}, \quad (1)$$

where V—volume air discharge, m^3/sec; T—its temperature, °K; ρ_a—density, kg/m^3 in normal conditions, P_1, P_2—pressure at the air hose input and output , respectively, Pa; z = 0.93—compression coefficient; a=0.01525м$^{1/3}$; R_a=287 J/(kg·K)—gas constant of air. To reach limiting operating characteristics of PHR motor with 50–150 m long hoses their diameter is to be not less than 11–13 mm. In the experimental sample of the tool we use 50 m long hoses with 15 mm flow section.

2.2. Unbalanced generator

Amplitude of vibration generator action force *F* depends on unbalanced mass rotation

$$F = m \cdot r \cdot (2\pi f)^2, \quad (2)$$

where *m*—unbalanced mass, m; *r*—its eccentricity, m; *f*—unbalanced mass rotation frequency (seismic vibrations frequency), Hz. It is necessary to regulate kinetic momentum of unbalanced vibrator *mr* to gain the preferable amplitude–frequency characteristics of the action force in 100–300 Hz frequency range. However, due to the limit of loading on bearing units and walls of an uncased borehole the action force maximum value in the developed source is limited to 15 kN. So, if the kinetic momentum of the unbalanced generator is not regulated, then, in case of vibration frequency drop from 300 to 100 Hz and in accordance with (2), the action force amplitude decreases by 9 times. It causes intensity drop below the permissible limit and reduction of the source frequency range amplitude. To solve the problem the vibration generator is based on the scheme of two unbalanced masses. One of the masses m_1 has constant eccentricity r_1, another one has m_2 which is $m_2 < m_1$, and its eccentricity may vary from r_2 to r_3 (Figure 2). Unbalanced mass gradual displacement along the borehole axis is caused by inertia force at the frequency f_i. It happens when the inertia force is superior to adhesive force Fr (f_i) of m_2 unbalanced mass with the axis on which the mass is fixed.

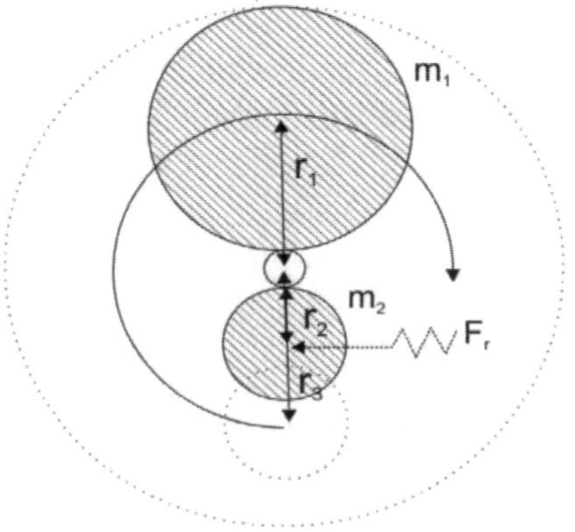

Figure 2. Scheme of a borehole vibration source vibration generator: m_1—main unbalanced mass; m_2—auxiliary unbalanced mass.

2.3. Pressure device

Pressure device is an obligatory part of a vibration source. It provides contact with the borehole walls and transmission of seismic vibrations to rock mass. Pressure device has to exceed the rated value of uncased borehole diameter by at least two times as in coal beds uncased boreholes diameter deviates from the design value. According to the work [11] a horizontal degassing borehole drilled with 98 mm diameter have a different operating diameter. It varies from 84 to 180 mm and varies greatly in all the parts of the borehole, especially at 10–20 m from the borehole collar. That is why anchor systems with small diameter margin such as wedge-like mechanisms are not applicable. The lever pressure device in the developed source is similar to the devices used in seismic tools for vertical seismic profile shooting of boreholes. The pressure device contains reversible DC motor. Through a multiphase planetary reduction gearbox and screw drive its rotation transforms into action of the lever, which moves forward and presses the vibration source to a borehole wall. Supply voltage of the pressure device is 24 W, current is not more than 1.5 A. The device can work in boreholes up to 240 mm in diameter. Control of the pressing force is done in accordance with ampere meter data from the remote control.

The main properties of the developed seismic impact facility:

—work in uncased 89–122 mm diameter boreholes with its tilting angle from -90 m to +90 degree;
—maximum diameter of boreholes 240 mm;
—action force amplitude up to 15 kN;
—impact frequency in hard rocks not less than 150 Hz;
—air pressure up to 0.8 MPa;
—air supply up to13.8 l/sec;
—virboseismic module length 1430 mm;
—16 kg mass.

3. Unbalanced source test

Testing has been carried out in loamy soil and solid hard rocks. In the first case the source has been placed at a depth of 3 m in a vertical uncased borehole 180 mm in diameter. The borehole is drilled in near-surface layer of mellow loam soil with the density of 1300 kg/m^3 density. In the series of experiments it is proved that a seismic P-wave velocity (V_p) in near-surface layer is 300 m/sec, in underlying soil—1350 m/sec. Seismic vibrations registration is performed by geophones GS20–DX (LLC OJIO–Geo limited, Oktyabrsk) set along the linear profile 145 m long at a step of 5 m. The signals are registered with the help of equipment ROSA–A (Siberian Research Institute of Geology, Geophysics and Mineral Resources, Novosibirsk). Air pressure in a pneumatic actuator of the source is changed from 0.4 to 0.7 MPa. We obtain that frequency of the first harmonic of seismic oscillations in soil depends on air pressure in the pneumatic actuator. The frequency is 35–36 Hz at the air pressure of 0.7 MPa, 30–31 Hz at 0.6 MPa and 24–26 Hz at 0.4–0.5 MPa. In the experimental conditions the actuator power is not sufficient to cause rotation of the generator unbalanced masses to the frequency more than 36 r/sec. The main reason is high reactive loading of the vibration source in the described medium [12].

Figure 3 shows the seismic energy flow spectra at several observation points at air pressure in pneumatic actuator of 0.7 MPa. Energy flow (seismic impact intensity) is estimated by the formula

$$\Phi = \frac{1}{2}\rho \cdot V_P \cdot \left(X_0'\right)^2, \quad (3)$$

where X_0'—velocity amplitude of rock displacement in vibration field. Value X_0' is calculated by the geophone output signals using the technique presented in the work [13].

The presented graphs show that the intensity of seismic impact drops fast in case of far distance from the source. Seismic energy flow at the basic oscillations mode decreases by 2 times with the distance increase from 1 to 5 m (Figures 3a and 3b).

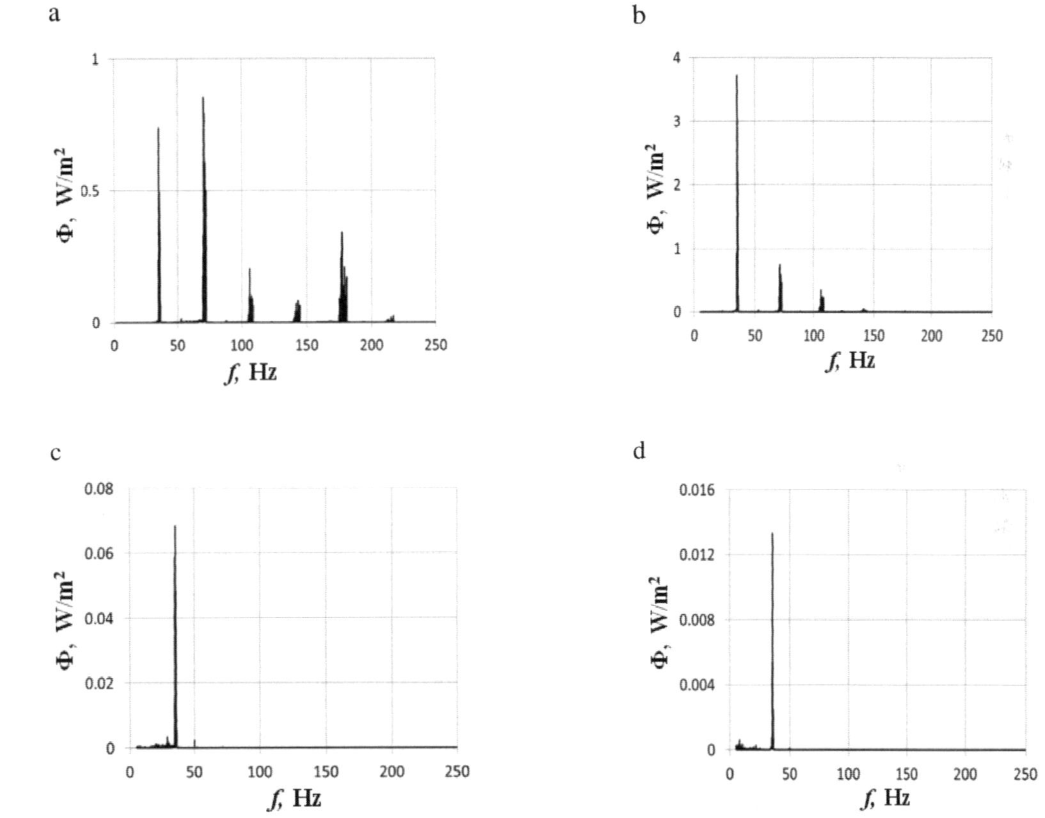

Figure 3. Seismic energy flow spectrum (seismic impact intensity) at various distances r from the source if the air pressure is 0.7 MPA: (a) r = 1 m, (b) r = 5 m, (c) r = 30 m, (d) r = 65 m.

The upper harmonic of oscillations decreases stronger, at the distance of 20–30 m they are negligibly small (Figures 3c and 3d). Compound flow of seismic energy in soils at the frequency of 250 Hz at the distance of 1 m from the source is equal to 8.8 W/ m^2, at 5 m—0.033 W/m^2, at 30 m—0.46 W/m^2. The radius of the vibration impact with the intensity higher than 0. 01 W/ m^2 is approximately 7 m. Taking into account basic vibration mode of 35–36 Hz frequency, the source impact radius does not exceed 3–4 m.

The source tests in solid rocks have been conducted in horizontal uncased boreholes with the diameter of 105 mm drilled in a pit wall. For the experiments pressure at the pneumatic actuator input has been changed from 0.6 to 0.8 MPa. Figure 4 shows the geophone output signal and amplitude spectrum of vibrations of bottomhole rock at a distance of 0.5 m from the source and at the actuator air pressure of 0.8 MPa.

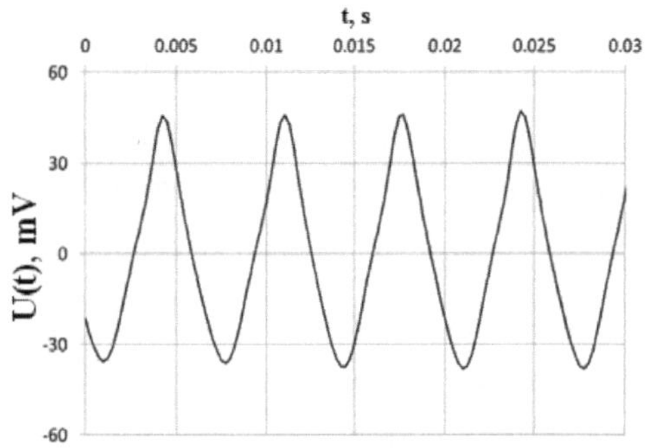

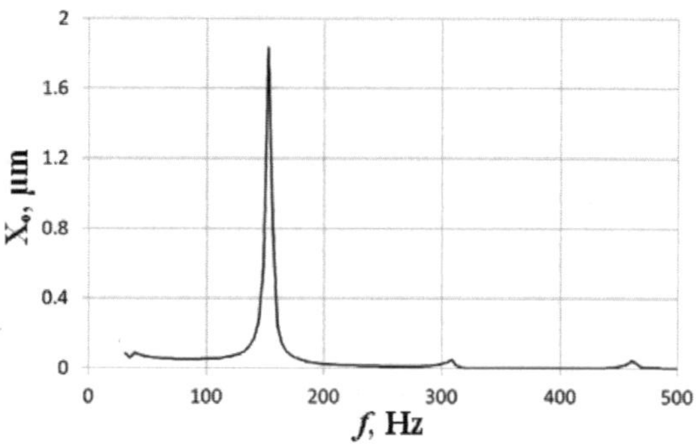

Figure 4. (a) Output signal of geophone and (b) and amplitude spectrum displacement of bottomhole rock mass in seismic field (r = 0.5 m).

The signal is similar to a sine wave (Figure 4a). Amplitudes of the second and third harmonics of seismic oscillations do not exceed the main mode of 152–156 Hz frequency (Figure 4b). Seismic energy flow in this frequency line is 19.2 W/m^2 at the observing point. The impact with the intensity more than 0.01 W/m has the radius of as 12 m.

4. Conclusion

The developed seismic source with a pneumatic actuator, electromechanical pressure device and unbalanced generator with automatic regulation of the kinetic momentum provides seismic impact on incased boreholes up to 240 mm in diameter with the action force amplitude up to 15 kN.

The source contributes to seismic impact on bottomhole of solid rock masses in a harmonic mode with the frequency up to 150–160 Hz and intensity more than 0.01 W/m^2 at the distance up to 12 m from the source.

In soft soils the source emits seismic signal the spectrum of which is rich with intensive upper harmonics. Due to the harmonics range of multifrequency processing of rock mass bottomhole at the distance of 3–4 m is widened up to 110–180 Hz. The radius of the basic mode vibrations impact on soils is 7 m with the frequency of 30–40 Hz and intensity more than 0.01 W/m^2.

References

[1] Westermark R and Brett JF 2003 Enhanced oil recovery with downhole vibration stimulation in Osage County Oklahoma (Oklahoma: Oil & Gas Consultants International, Inc.) doi: 10.2172/822922

[2] Kurlenya MV and Serdyukov SV 1999 Determination of vibroseismic area affecting oil deposits from daylight surface *J. Min. Sci.* Vol 35 No 4 pp 333–40

[3] Serdyukov SV and Kurlenya MV 2007 Mechanism of oil production stimulation by low-intensity seismic fields *Acoust. Phys.* Vol 53 No 5 pp 618–628

[4] Serdyukov SV and Kurlenya MV 2007 Seismic stimulation of oil reservoirs *Russ. Geol. Geophys.* Vol 48 No 11 pp 897–901.

[5] Skazka VV, Serdyukov SV and Kurlenya MV 2014 Downhole unbalance vibration exciter near-field analysis *J. Min. Sci.* Vol 50 No 6 pp 1026–32

[6] Skazka VV and Serdyukov SV, Erohin GN and Serdyukov AS 2013 Near-field range of the

direct-impact seismic source *J. Min. Sci.* Vol 49 No 1 pp 60–7
[7] Rybalkin LA, Skazka VV and Serdyukov SV 2015 Equipment and procedures of downhole vibration stimulation of gas drainage *Interexpo GEO-Siberia* 2 (3) pp 216–21
[8] Timonin VV and Kondratenko AS 2015 Transportation system of technological and measuring equipment in uncased boreholes *J. Min. Sci.* Vol 51 No 5 pp 1056–61
[9] Chichinin IS 1984 *Seismic Waves Vibrational Emission* Moscow: Nedra (in Russian)
[10] Chuhareva NV 2010 *Simple and Complex Gas-Pipe Lines Calculations* Tomsk: TPU (in Russian)
[11] Mills K, Jeffrey R, Black D et al 2006 Developing methods for placing sand–propped hydraulic fractures for gas drainage in the bulli seam *Underground Coal Operators Conference*. Wollongong, Australia pp 190–9
[12] Borzov VM and Ivlev VI 2009 Development of operational characteristics of a laminal pneumatic motor due to improved properties of engineering materials *BHTP Vestnik* No 25 pp 2–6
[13] Serdyukov SV, Kurlenya MV, Patutin AV, Rybalkin LA and Shilova TV 2016 Experimental Test of Directional Hydraulic Fracturing Technique *J. Min. Sci.* Vol 52 No 4

Main development trends and some technical decisions on mining tools equipped with super-hard composite materials inserts

LT Dvornikov[1], PD Krestovozdvizhensky[2], SM Nikitenko[2], VA Korneyev[1] and PA Korneyev[1]

[1]Federal State Educational Institution of Higher Professional Education, Siberian State Industrial University, Novokuznetsk, Russia
[2]Federal Research Center for Coal and Coal Chemistry, Siberian Branch of the Russian Academy of Sciences, Kemerovo, Russia

E-mail: [1]tmmiok@yandex.ru

Abstract. Directions of a significant increase in effectiveness of rock destruction with tools equipped with super-hard composite material inserts are reviewed and justified. Drilling tool designs with the cutting insert in the form of elliptical Cassinian oval and the asymmetric ring cleaves are suggested. Versions of laboratory stand designs in order to determine the power consumption of rock destruction are developed.

Rock crushing is the main process of mineral deposit development. The mechanical crushing method widely used for this purpose at mining enterprises needs creation of high strength wear-resistant tools which can stand increasing power loads conditioned by steady increase in drive power of rock crushing machines. Technical solutions embodied in the design of the existing mining tools, most of them have been proposed in the second half of the twentieth century and are in need of modernization in accordance with the results of scientific and technical progress.

The authors conducted theoretical studies in order to justify new designs of rock crushing tools, reinforcement materials for their cutting inserts, and also tested the experimental samples of mining tools within the project framework of the Federal Target Program on Development of Experimental Combined Tool Designs Using Super-Hard Composites for Effective Crushing of Rocks.

The results received allow identifying the most promising modernization trends of the existing mining tools nomenclature that enhance their productivity, wear resistance, and also reduce specific energy consumption in the process of mine rock crushing. Rock crushing tool efficiency increase can be achieved by various ways: the tool geometry change, the use of new materials in the manufacture of cutting inserts, as well as the use of new, more productive and energy efficient modes of rock crushing.

The first direction includes research on the development of various types of mining tool structures providing guaranteed turnability of revolving cutters, prevention of rear cutter face from lowering, more rational armoring of bladeless tool, as well as radial cutters use range expanding according to the rocks strength without changing the cutting insert geometry.

Based on the results of the fulfilled researches new drill bits designs are developed: with a cutting insert of Cassinian oval elliptical shaped [1] and also with asymmetric ring positioning [2]. The cutter with a cutting insert of elliptical Cassinian oval shape is shown in Figure 1a. It consists of a body 1, two styluses with hard alloy inserts 2 of elliptical Cassinian oval shape, blades 3 set along the big oval axis wherein the major axes of the ovals are inclined to the tool axis at an acute angle α.

One of the properties of an elliptical Cassinian oval is the presence of two points with a curvature tending to infinity, which allows the full contact when interacting with the face at these points. This feature of the cutter helps to reduce cutter blades blunting 3, in contrast to the existing cutters, in which the interaction of the blades with the face is carried out along the stress concentrators placed on them.

In order to reduce the energy consumption of rock drilling, by providing their destruction with major cut-offs within the frames of the study the drilling bit with asymmetric ring positioning was designed (Figure 1b). The drilling bit was created on the basis of RKS bit design, consisting of a body and discontinuous blades placed symmetrically to the bit axis [3]. The disadvantage of the RKS tool is the fact that both inner semi-blades cut on the face of the borehole concentric strips, easily imposed on each other, which does not lead to active face destruction between the cutting strips, i.e. power consumption of destruction process is high.

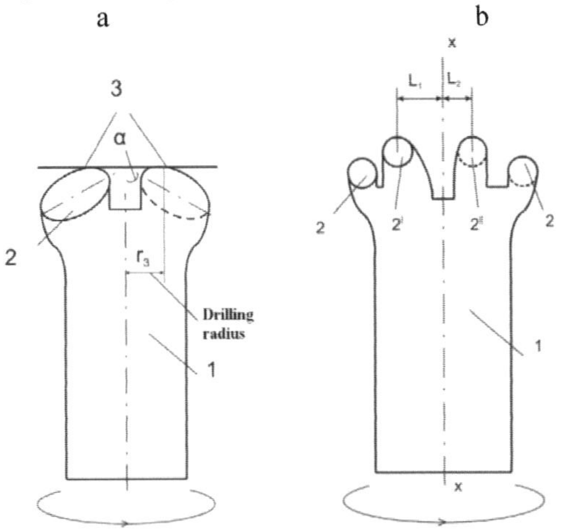

Figure 1. Drilling bits with the cutting insert of elliptical Cassinian oval shape (a) and asymmetric ring positioning (b).

Considering this drawback of PKS cutter, in the design of the cutter with asymmetric ring positioning the inner semi-blades are located asymmetrically with respect to the axis of the cutter rotation (Figure 1b). The developed cutter design with asymmetrical ring positioning allows different semi-blade combinations depending on borehole drilling conditions.

Drilling tool with asymmetrical ring positioning (Figure 1b) consists of a body 1, two styluses with discontinuous blades constructed as peripheral 2 and the internal 2' and 2" semi-blades while internal semi-blades 2' and 2" are set asymmetrically with respect to the cutter axis, at the distances L1 and L2. During drilling, the cutter is rotated according to the arrow in Figure 1b with simultaneous axial advance to the face. Thus semi-blades 2, 2' and 2" start cutting rock from the face. Due to the asymmetric installation of the internal semi-blades 2' and 2" the cut strips overlap each other thus providing the drilled borehole face destruction with big pieces, resulting in reduced power consumption of the drilling process.

As it was previously noted, when considering ways to improving efficiency of destruction of rocks by drilling cutters, one of the most urgent problems to be solved during the design process is to prevent the collapse of the rock face with the cutter rear side.

Theoretical studies carried out in the framework of the Federal Target Program allowed justifying fundamentally new designs of drilling bit cutting blades.

Figure 2 depicts the bit movement geometry, performing rotational movement around the axis $0^{|}0$ with the speed ω and the simultaneous forward advance to the depth h per one revolution.

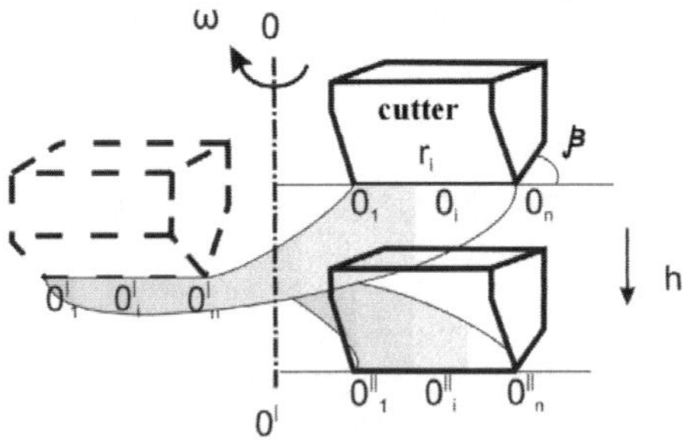

Figure 2. Drilling bit movement geometry.

It is seen in Figure 2 that each point of the cutter blade 0_1, 0_i, … 0_n remote from the tool axis at a distance r_i, passes the path $0_10^{|}_10^{||}_1$, $0_i0^{|}_i0^{||}_i$, … $0_n0^{|}_n0^{||}_n$ during the cutter turnover, respectively.

Helical trajectory scan of each point 0_1, 0_i, … 0_n in Figure 2 represents the hypotenuse S of a right triangle the legs of which are the distances of the drilling tool advance h and the circumference length with r_i radius scribed with a corresponding point relatively to the axis $0^{|}0$ (Figure 3a). In this way the trajectory of each cutter blade point is the angle of inclination α_i with the cutting plane of the cutter at zero advance (Figure 3a). This angle can be found as:

The value of the angle α_i is variable and decreases from its inner edge to the outer edge. This fact imposes certain restrictions on the clearance value of the rear cutters angle β since achieving equality α_i=β drilling cutter stops cutting, causing the collapse of the rock.

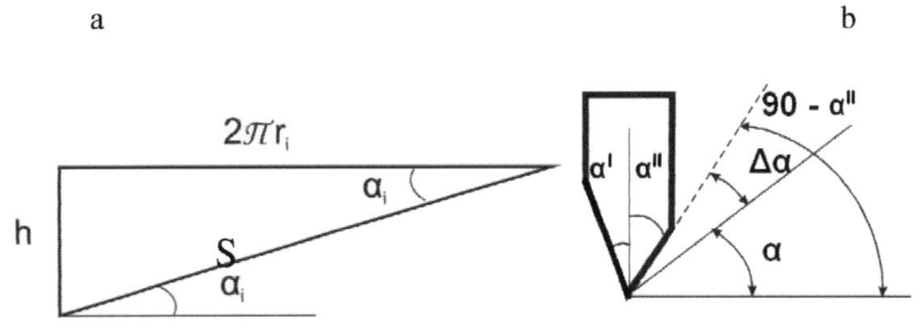

Figure 3. Helical trajectory scan of the cutter point (a) and cutting geometry (b).

Usually, in practice [3], mining tool rear angle is within the 5–20°. With positive front angle rear angle increase leads to weakening of the cutting edge of the tool and may lead to its failure. With negative front angle, as the studies have shown, rear angle can be increased up to 20–30°, as the strength of the cutting edge is not reduced, and the wear area have less values.

It is obvious that the cutter cutting edge strength depends directly on the sharpening of its angle which is composed of the front $\alpha^{|}$ and rear $\alpha^{||}$ sharpening angles (Figure 3b). At the same time, to avoid rock collapsing with the rear cutter blade, it is required that Δα equal the difference between the cutting angle α and the rear sharpening angle in accordance with condition (1)

$$\Delta\alpha = (90 - \alpha^{||}) - \alpha . \qquad (1)$$

The cutter design corresponding to condition (1) can be obtained when using variable values of back sharpening angle $\alpha^{||}$ over the entire length of the blade [4]. Considering the plots in Figure 3 and using drilling cutter advance meaning h, we can calculate the sharpening angle value $\alpha^{||}$ for any point of the blade (Figure 4).

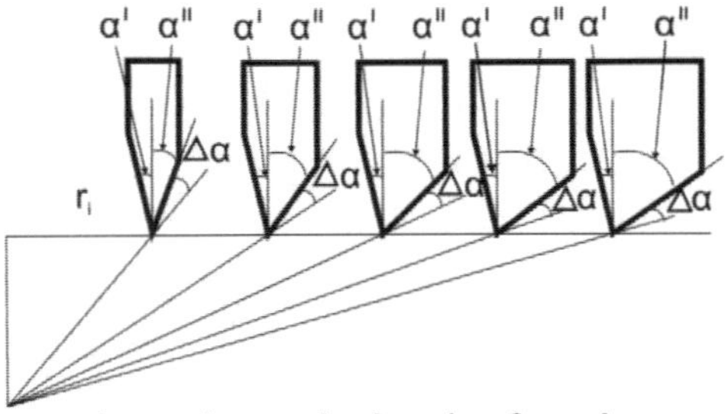

Figure 4. The rear cutter sharpening angle changing from the center to periphery.

Thus, the implementation of variable adjustable sharpening angle α^{II} of the cutting edge across the length of the blade in the design of the tool helps to prevent collapse of rocks with the cutter back blade by providing difference Δα between the cutting angle α and rear sharpening angle.

Reducing the sharpening angle from the periphery to the center can be justified by the fact that the moment of cutting resistance at the points of the tool from the periphery to the center is reduced due to reducing of these points radii with respect to the tool axis.

Apart from the considered case, avoiding rock crushing with the cutting tool back blade can be implemented in the design of the drilling bit with constant blade sharpening angle under the following condition

$$\alpha^{\text{I}} + \alpha^{\text{II}} = \alpha^{\text{III}} = const, \qquad (2)$$

where α^{III}— sharpening angle of the bit cutting insert.

Cutting insert design [5] corresponding to the conditions (1) and (2) is realized through the provision of variable rear and front angles of sharpening in compliance with the constancy of the cutter sharpening angle (Figure 5). The advantage of this design is the same strength of the cutting insert along the entire length of the blade.

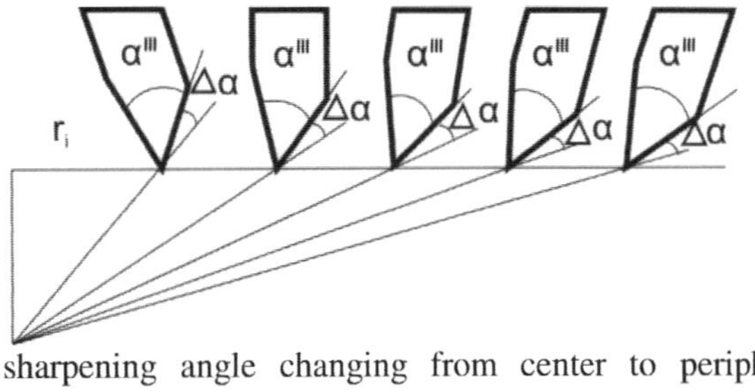

Figure 5. Cutter sharpening angle changing from center to periphery preserving sharpening angle constancy.

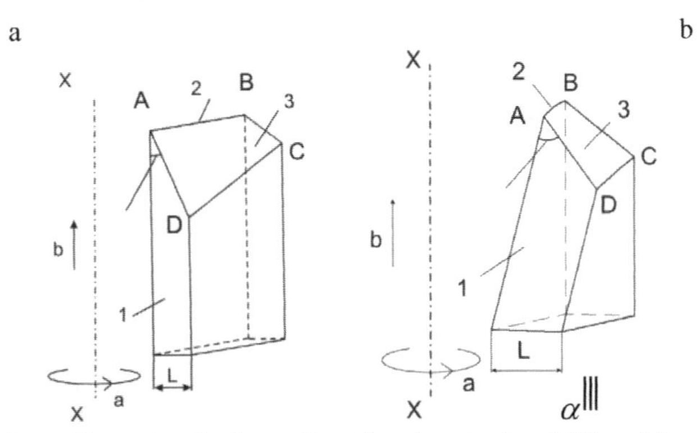

Figure 6. A three-dimensional view of cutting inserts for drilling bit with a variable sharpening angle (a) and a constant sharpening angle (b).

 IOP Publishing
doi:10.1088/1755-1315/53/1/012031

Three-dimensional view of the cutting inserts with variable and constant sharpening angles is shown in Figures 6a and 6b, respectively. Figure 6 depicts: 1—insert body, 2—blade, 3—rear insert edge, X–X geometric drilled borehole axis, arrows a and b show the insert rotation direction with respect to X–X and its advance—deepening of the cutting insert into the produced borehole. Points A, B, C, D, are used to indicate the tops of the insert, forming its rear edge, L—thickness of the cutting insert, α^{III}—the angle of its sharpening.

During the research under the Federal Target Program, possible ways to increasing efficiency of rock crushing by changing the indenter geometry and layouts in the edgeless drill bits were also reviewed. It is known that edgeless cylindrical spherical hard alloy inserts are usually located at the end of the bits according to different layouts which quite often do not correspond to the rock crushed and the face geometry. Thus, an excessive concentration of the inserts in the geometric center of the face can result in rock overgrinding which in its turn leads to increased power consumption for drilling.

One of the options for ensuring the reduction of drilling energy consumption may be execution of a concentric recess on the bit casing striking surface (Figure 7). Such drill bit construction [6] consists of a casing 1, its striking surface 2, indenters 3 of the given diameter d with the given radius r of their working bulge h. On the striking surface 2 a concentric recess 4 is made with depth h_1 not less than the indenter working bulge h ($h_1 \geq h$), and the concentric recess 4 diameter d_1 is at least twice the diameter of the indenter d, ($d_1 \geq 2d$).

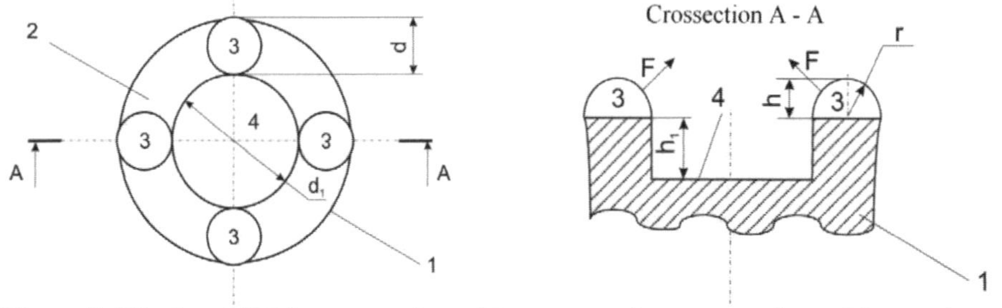

Figure 7. Edgeless drill bit construction with a concentric recess on the striking surface.

When such bit impacts the borehole face and because its striking surface 2 has a concentric recess 4, a core is formed. The formed core while drilling further is easily sheared off due to the efforts from the indenters located on the edge of concentric recess 4, because of the occurring forces F, directed normally to the surface of the indenters.

All currently used edgeless inserts are built on the principle of the surface curvature constancy of the insert in the area of its interaction with the rock, i.e. the directly impacting surface is manufactured of a spherical shape. The disadvantage of this design is the fact that when impacting the rocks such indenters have no directional effect, i.e. destruction can occur equally in all directions from the point of contact. To have the opportunity to manage the process of destruction in a preferred direction, it is advisable to create indenters with different curvatures in main directions.

One of such insert variant [7] is shown in Figure 8: 1—the body of the insert; 2—the working portion of the insert; P—the insert working portion surface; M—the contact point of the insert working area P surface with the rock; K—contacting P surface at M point the tangent plane—the plane of the destruction face.

It is known from the theory of surfaces that through the surface and the tangent plane contact point one can draw two main direction lines I–I and II–II, which, being mutually perpendicular, define in their cross sections the largest k_1 and the smallest k_2 of curvature, i.e. in section I-I there appears plane curve with a radius of curvature ρ_1, and in section II–II—plane curve with a radius of curvature ρ_2. In this case the relations $\frac{1}{\rho_1} = k_1$ and $\frac{1}{\rho_2} = k_2$ become curvatures.

The suggested insert for drill bits is manufactured with different main curvatures ($k_1 \neq k_2$), i.e. at the insert cross-section with the plane perpendicular to the axis y-y insert elliptic curve appears with different geometric axes. In Figure 8 an example of a special case of such insert is shown for the ratio $\frac{k_1}{k_2} = 2$.

The suggested insert impact principle on a face consists in the fact that the axial force acting in different directions inside the main directions creates various states of stress in the crushed medium and leads to the destruction along the axis of the greatest curvature thus controlling the destruction process.

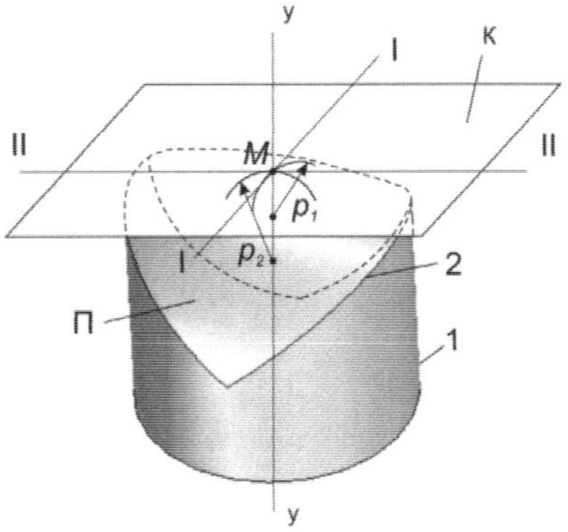

Figure 8. Insert geometry for an edgeless drill bit which provides the mine rock crushing process control.

In addition to using effective geometry in the construction of the tool blade and body, it becomes important for the mining tool performance improvement to provide the tool cutting inserts with better wear resistance. Thus, the results of the statistical analysis of drill and combine cutters wear types, conducted in 2014 in the first phase of the Federal Target Program implementation, show that the vast majority of the tool failure causes happen due to the blunting of the cutting inserts.

One of the most promising ways of improving wear resistance of the rock cutting tool is the use of inserts of super -hard composite materials, which are soldered or mechanically fastened on its bearing steel part. The polycrystalline super-hard materials (PSTM) are industrially manufactured in large quantities and are divided into two big groups: made on the basis of cubic boron nitride and diamond-based (Table 1). Polycrystalline cubic boron nitride PCBN is obtained by reactive sintering in a high-pressure apparatus (HPA) under pressures from 4.0 to 6.0 GPa and temperatures from 1300 to 1500°C. Cubic boron nitride is characterized by quite high hardness and, compared with diamond composite materials, possesses excellent chemical resistance to materials containing iron, nickel, and carbon. For this reason PCBN is more widely used in the metal-working industry as a cutting tool for continuous and impact sharpening of quenched steels and irons. The wear-resistance of the composite PCBN depends first and foremost on the content of cubic boron nitride in the composite as well as on the composition and heat resistance of the bonding phase. Tool wear at the cutting boundary is due to partial recrystallization of the material and embrittlement of the cutting edge owing to the interaction with the worked material.

Composites based on polycrystalline diamond PCD, which are obtained by thermobaric synthesis, are widely used for working non-ferrous metals and their alloys as well as in areas associated with the development of deposits and mining minerals. Owing to high hardness diamond composites are used predominately in operations where the main mechanism of tool wear is abrasive wear. However, polycrystalline diamond has a number of limitations in operation that significantly degrade the tool efficiency. One such limitation is the impossibility of using this composite at high temperatures (> 400°C).

Table 1. Comparison of the physical properties of different composites based on polycrystalline diamond, synthetic diamond single crystal and hard tungsten–cobalt alloy.

Index	PCD (PDC)	Heat resistant PCD (TSPD)	Synthetic single crystal	Sintered hard alloy VK15
Density, g/cm3	3.90–4.10	3.42–3.46	3.51–3.52	13.80–14.10
Microhardness, GPa	50–60	50–55	70–105	15–18
Strength in compression, GPa	7.5–8.5	4.0–5.5	7.0–9.0	12.0–14.0
Fracture toughness, MPa $m^{1/2}$	9.0–10.0	7.0–8.0	3.0–4.0	11.0–12.0
Young's modulus, GPa	810–850	920–960	1000–1150	70–100
Thermal conductivity, W m^{-1} K^{-1}	550–750	120–250	700–1500	70–100
Wear resistance, mm	0.15–0.20	0.20–0.25	–	> 2.0
Heat resistance, °C	700	1200	1200	1400

The first reason for this is structural degradation owing to the difference of the linear thermal expansion coefficients (CLTE) of the metal or carbide bonding phase and the base diamond phase. It is well known that increasing the operating temperature of the tool to 400°C and the different CLTE of the components of the composite material promote partial breaking of the diamond–diamond bond at the interphase boundary, which results in crack formation and subsequent chipping of the cutting edge of the PCD composite during operation.

The second reason is the thermal degradation of the structure during operation owing to the presence of bonding materials, viz. metallic cobalt, which is classic for PCD, in the intermediate layers of the composite and gradual dissolution of the diamond grains in the cobalt at heightened temperatures.

Different additional treatments of sintered materials (etching cobalt out of the near surface layer) as well as the development of fundamentally new recipes (systems with close CLTE, for example, diamond– silicon carbide) and hybrid materials (for example, layered composites of the type WC–Co–diamond–SiC) are being proposed in order to solve the problems of the low heat resistance of classic PCD composites.

Today, bilayer PCD composites obtained under pressures 6–8 GPa and temperatures 1400–1600°C are widely used for resource exploration and mining (drilling industry). A diamond-containing layer is formed owing to the bonding material Co–WC–C, which infiltrates from the substrate based on sintered tungsten carbide. The high mass content of cobalt in the diamond layer (from 6 to 10%) greatly lowers the heat resistance of the diamond insert. Owing to an increase of the cutting speed the temperature on the cutting edge of the composite increases to 700°C, which results in catastrophic thermal and chemical wear of PCD inserts.

Despite positive experience of super-hard materials use, the biggest market share in the cutting insert constructions based on tungsten cobalt hard alloys. One of the main reasons for this is a significant cost of super-hard materials, which for cubic boron nitride is from 1000 USD per kilogram, against USD 100 per kilogram for classic hard alloy. With this in mind, the use of super-hard materials must be selective, that is, they can be used as reinforced inserts of the of hard alloy blades.

In this case it is important to justify the most critical sections of the blades, the installation in which the super-hard elements can be justified in terms of the required crushing performance and the cutters wear resistance. Thus, the rotary drilling field expansion in rocks with the strength of 100–120 MPa is possible through the use of the combined cutting inserts equipped with elements made of super-hard materials of PCBN or PCD type in the design of cutting blades. The advantage of such designs is the selective reinforcement of inserts, allowing, on the one hand, using a super-hard material exclusively in areas of high stress concentration, and on the other considerably save expensive composites.

One of the variants of a cutter design with combined cutting inserts [8] developed with the implementation of applied scientific research, is shown in Figure 9. It consists of a cutter body 1, blades 2 and cutting inserts 3. Wherein the inner 4'and outer 4"with respect to the cutter rotation axis x–x trihedral corners of the cutting inserts, are the inserts stress concentrators, and are made in the form of super-hard material element—PCBN or PCD.

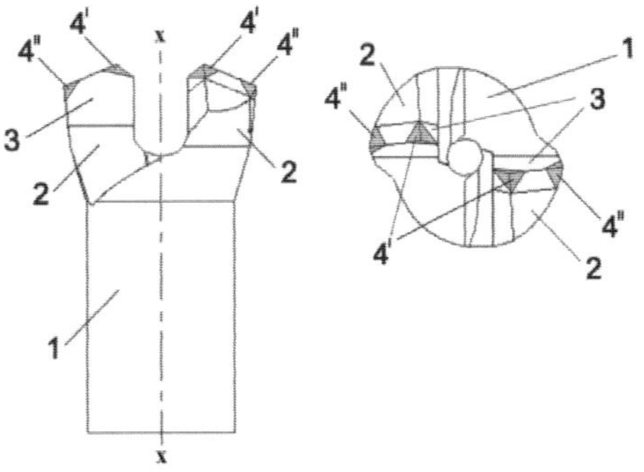

Figure 9. Drilling cutter with combined cutting inserts made of tungsten cobalt super-hard alloy and composite materials.

The idea of selective reinforcement of rock cutting tool inserts with super-hard materials also found its application in the design of experimental configurations of tangential rotary cutters (Figure 10). The developed cutting insert [9] is a two-stage form structure with the fixing stage, designed as a cone with self-retarding corners, and a cutting step performed in axial section according to the law of the power function y = xn with n = 2. Cutting insert surface is reinforced with indenters made of composite wear-resistant material disposed in the plane along the spiral of Archimedes. The proposed design of the cutting insert will provide a definite cutter turning effect, due to the asymmetry of the load application.

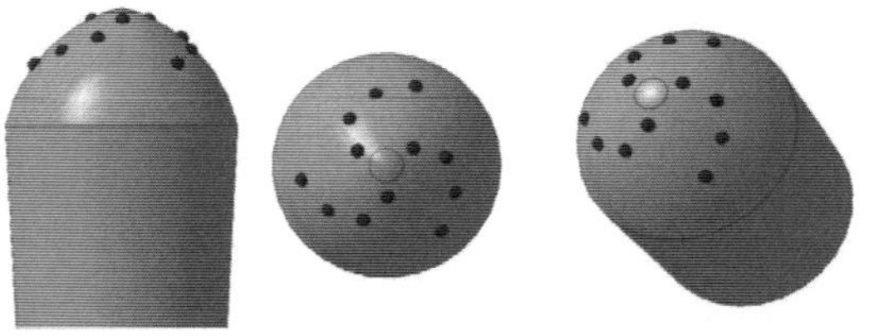

Figure 10. Various options for the location of super-hard materials of the tangential rotary cutter cutting insert surface made of tungsten cobalt alloy.

Rock crushing efficiency increase can also be achieved through the use of new, more productive and energy efficient modes of rock crushing. With respect to the rotary hole drilling it is a fundamental improvement of the drilling method for the purpose of its use in rocks of medium and above medium strength, i.e. in rocks with the strength factor > 10, where rotary drilling method is traditionally considered unsuitable due to insufficient strength of cutters and first of all because of the low wear resistance of the cutters. To expand the scope of rotary drilling method and for drilling speed increase super-hard materials must be used for the cutting insert reinforcement, and also more powerful drilling machines must be created.

The task of a substantial (by a few times) increase in drilling productivity, which means the drilling rate, can be expressed as follows

$$V = h \cdot n, \qquad (3)$$

where V—drilling rate; h—advance per one bit rotation; n—number of bit rotations.

It is known that the relationship between the drilling rate *V* and the tool number of revolutions *n* is not linear in view of the optimal tool rotation frequency for rocks with different strength factor [3]. Accordingly, the drilling performance resource may be found only in the cutter advance increase, resulting in a significant increase of a cutting angle.

Table 2 shows the calculated data on the cutting angle changing depending on the cutter point radius remoteness from the blade axis and the length of its advance. A graphical representation of the calculated data is shown in Figure 11: it is seen that the cutting angle decreases from the inner edge to the periphery, while with the advance increasing the values increase. Thus, the increased advance drilling requires the use of cutters with reduce the cutting-point angle in order to avoid touching the crushed rock with the back edge of the cutter. Decreasing the cutting-point angle of tools reduces their strength under the existing blade manufacturing technology, which in turn imposes limitations on the increase in the cutter advance rate. The design developed in the framework of the conducted applied research and shown in Figure 6b can serve as the way out from this situation. A characteristic feature of this design is the presence of variable front and rear cutting-point angles.

Table 2. Cutting angle dependence on blade point remoteness radius from the cutter axis and the length of its advance.

Advance, mm	Blade point remoteness radius, mm								
	5	7	9	11	13	15	17	19	21
3	5.46	3.91	3.04	2.49	2.11	1.83	1.61	1,44	1.30
4	7.26	5.20	4.05	3.32	2.81	2.43	2.15	1.92	1.74
5	9.05	6.49	5.06	4.14	3.51	3.04	2.68	2.40	2.17
6	10.82	7.78	6.06	4.97	4.21	3.65	3.22	2.88	2.61
7	12.57	9.05	7.06	5.79	4.90	4.25	3.75	3.36	3.04
8	14.30	10.32	8.06	6.61	5.60	4.86	4.29	3.84	3.47
9	16.00	11.58	9.05	7.43	6.29	5.46	4.82	4.32	3.91
10	17.67	12.82	10.04	8.24	6.99	6.06	5.35	4.79	4.34

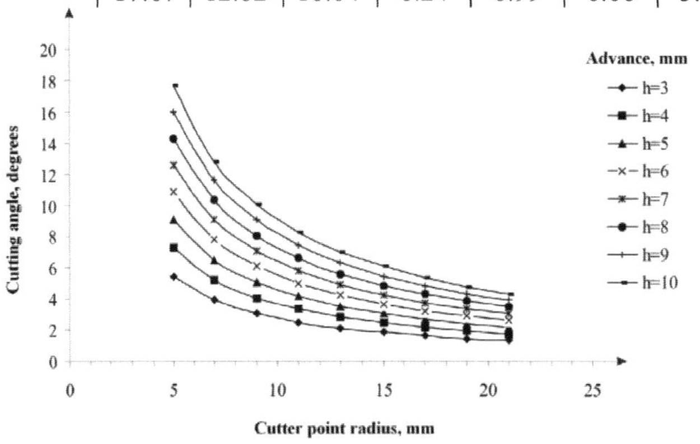

Figure 11. The angle of cutting blade points depending on the distance from the drilling cutter rotation axis for different advance values.

An additional advantage of drilling with the increased advance is to reduce the rock abrasion impact on the cutter due to reduction of the S path traveled by points placed on the cutter. In Figure 3a it is seen that the path S traversed by the cutter blade point depends on the size of the specific advance. The total path travelled by the cutter blade point in n revolutions, can be found as

$$S_{\sum} = S \cdot n, \qquad (4$$

where $S_{\sum}$ —the total path travelled by the cutter blade point; *S*—the path traversed by the cutter point at one revolution; *n*—number of the cutter's revolutions.

It is obvious that during drilling a drill with higher specific advance h needs to make less cutter revolutions n when drilling to the same depth. Then, according to the relation (4) reduction of the cutter revolution number n will reduce the total path traveled by the cutter blade point.

It should be noted that physical phenomena accompanying the drilling process impose limits both to specific advance and to the tool revolution frequency. In addition, the parameters h and n are interdependent values, i.e. n change causes naturally change of h. When drilling boreholes by rotary method using the existing machines a specific advance h can vary from zero up to 10–15 mm/rev. Therefore, these values can be considered achievable in principle. At the same time there is objective limit for advance h due to the tool blades limited strength.

Such restrictions are reduced to the limit values of axial forces and torques, which can withstand the cutters, without being destroyed. The strength of the existing drilling tools is limited by axial forces P = 15–20 kN. When you create a more durable material for cutting tools and increase axial forces up to 30 kN and above, you can expect a significant increase in drilling performance. Generalizations show that by using the existing tools one can stably provide averaged values of h = 2.5–4.5 mm/rev. The presence of cracks and interstices in rock mass necessitates control of specific advance for the purpose of the tool protection against overloads and failures (it is the simplest way to provide constant specific advance of the tool).

Attention should be also paid to the rotary drilling method dynamic features. The frequency of the drilling rod rotation and its torsional stiffness directly affects the speed of drilling. Specially conducted experiments showed that the rate of drilling with the same axial force and the same rotational frequency changes depending on the rigidity of the rods. Thus, after 20 times reduction of rod rigidity advance rate is reduced by almost half.

Under real conditions of borehole drilling rod rigidness may change more than 10 times, and during well drilling—100 times. The torsional stiffness of the rods directly affects the thickness of the removable chips. The smaller is the rigidity of rod, the smaller is the removable chip thickness.

In addition, it can be affirmed that the rock like a fragile medium under crushing in the mode of alternating chips is the source of vibrations in the rotating machine–rod–cutter system. Vibration parameters are determined by the elastic characteristic of the system and above all by the drill rod characteristic. Rod rigidness increase leads to higher frequency vibrations with small amplitudes without actually and finally affecting the drilling process result. The increased torsional stiffness of the rods and transmissions as well as reduced inertia of the advancing mechanism positively influences drilling efficiency.

References

[1] Dvornikov LT, Nikitenko SM, Korneyev VA and Korneyev PA 2015 Drilling cutter with a cutting insert in the form of Cassini elliptical oval *Patent Application 2015125831*

[2] Dvornikov LT, Nikitenko SM, Korneyev VA and Korneyev PA 2015 Drilling cutter *Patent Application 2015125832*

[3] Alimov OD and Dvornikov LT 1976 *Drilling Machines* (in Russian)

[4] Dvornikov LT, Nikitenko SM, Korneyev VA and Korneyev PA 2016 The drilling tool cutting insert *Patent Application 2016103526*

[5] Dvornikov LT, Nikitenko SM, Korneyev VA and Korneyev PA 2016 The drilling tool cutting insert *Patent Application 2016103561*

[6] Dvornikov LT, Nikitenko SM, Jukov IA and Korneyev VA 2016 Drill bit *Patent Application 2016145061*

[7] Dvornikov LT, Nikitenko SM, Jukov IA and Korneyev VA 2016 Carbide inserts for the drill bit *Patent Application 2016145024*

[8] Dvornikov LT, Nikitenko SM, Korneyev VA and Korneyev PA 2016 Drilling tool for rotary drilling of holes *Patent Application 2016145026*

[9] Dvornikov LT, Nikitenko SM and Krestovozdvizhensky PD 2016 The cutting insert of a tangential rotary cutter *Patent Application 2016125441*

IOP Publishing
doi:10.1088/1755-1315/53/1/012032

Geological and technological evaluation of gold-bearing mineral material after photo-electrochemical activation leaching

DV Manzyrev

Chita Division, Chinakal Institute of Mining, Siberian Branch, Russian Academy of Sciences, Chita, Russia

E-mail: geoxxi@mail.ru

Abstract. The paper reports the lab test results on simulation of heap leaching of unoxidized rebellious ore extracted from deep levels of Pogromnoe open pit mine, with different flowsheets and photo-electrochemically activated solutions. It has been found that pre-treatment of rebellious ore particles –10 mm in size by photo-electrochemically activated solutions at the stage preceding agglomeration with the use of rich cyanide solutions enhances gold recovery by 6%.

1. Introduction

A life cycle of a gold mine using heap leaching in Transbaikal area consists of two bold stages. The first stage is extraction of gold from oxidized ores of the upper horizons. The second stage operations embrace low-grade proto-ore of the lower horizons, which results in reduction in recovery rate by tens percent. For this reason, purposely to sustain economically sound extraction of god from rebellious ore, the innovative approaches to ore pre-treatment and leaching are required, with the parameters and regimes adequately supported by geological and technological evaluations.

The aim of this research was to assess influence of photo-and-electrically activated solutions used in ore pre-treatment and agglomeration on the ratio of extraction of useful component from proto-ore of Pogromnoe deposit.

2. Research

Pogromnoe deposit occurs in the Shilka district of the Transbaikalia, 15 km eastward of Shilka town. The deposit adjoins a scute of the complex-structure overthrust of the Mongolia–Okhotsk suture. The overthrust scute is composed of Triassic–Low Jurassic igneous–sedimentary rocks, with sandstone and siltstone at the bottom and rhyolite clastic flow at the top of the formation. Horizontally, the scute is 1–1.5 km thick. The scute is rhomb-lens-shaped with the northeastward extension for 5–6 km. During active overthrusting, the ore-hosting rocks were heavily altered and transformed into clastic rocks. Metasomatic processes are widely developed in rocks, up to formation of secondary quartzite [1].

Gold-bearing ore of the deposit belongs to the gold–moderately sulfide–quartz formation represented by two morphological types: ore-fold quartz–carbonate–arsenopyrite–pyrite species in metasomatically altered effusive blanket (ore body 1) and strong-and-vein quartz species (with pockets of sulfides) in altered carbon-bearing shale (ore body 10). Material constitution of the ore is to 85–90% quartz, sericite, albite and carbonate, and 10–15% of the composition is ore with the main components of pyrite and arsenopyrite. The accessory minerals (tenth of a percent) are mainly sphalerite, chalcopyrite and pyrrhotine. The most typical textures are dissemination, veins, les often spots and strings. Cataclastic textures, breccias, jointing and schistosity are abundant. In ore body 1, most of gold particles (75–80%) have size less than 16 μm. Gold is free and mainly high-grade. Ore body 10 contains different size gold grains, from fine and dispersed to coarse (larger than 0.25 mm). Also, gold is free and high or very high grade [2].

The Transbaikal Integrated Research Institute has revealed two commercial types of ore: type I is gold–albite–sugary grained quartz with or without arsenopyrite (pyrite), jatosite and scorodite; this is the oxidized ore of the residuum, making up 85–90% of the in-place reserves; type II is gold–sugary grained quartz–poor sulfide proto-ore amounting to 10–15% in the in-place reserves. Type I ore of the upper horizons at least down to a depth of 50 m is assumed as easy-to-concentrate. Based on that, the Transbaikal Integrated Research Institute has proposed to use the method of heap leaching after two-stage crushing to a size of -20 mm. From the laboratory testing data, gold recovery in this case will make up 71.4–91% depending on ore porosity [3].

In 2005 Aprelkovo Mine commenced mining and heap leaching of Pogromnoe ore. In 2007, the Mine reached the design capacity at annual gold-bearing ore production of 850–1000 kg. The average content of the useful component was 1.2 g/t at the standard deviation of 0.19 g/t. However, gold recovery was 60% up to 2012 and reduced to 47–48% in 2012–2015.

The reduction in the gold recovery was connected with the commencement of mining and processing of type II unoxidized ore. From the evidence of the TOMS Institute, the mineral composition of this ore contains 96.5% of rock-forming minerals, including quartz, feldspar, micaceous minerals and carbonates. Ore minerals revealed in a sample in a quantity of 2.3% are all sulfides. More than 65% of sulfides is pyrrhotine, the rest are equal amounts of arsenopyrite and pyrite. Sphalerite is met as units. Accessory minerals total 1.2% in the gross mass of the ore [4].

For percolation filtration tests, samples of gold–quartz–poor sulfide proto-ore with the size of particles of –10 mm had 17.4% content of size grade +10 mm. Bulk weight of the test material was 1544.7 kg/m^3, moisture content—17.79%.

For cyanidation in the laboratory percolation filters, the initial samples were washed off slime that was cyanated in bottle conditioners. The content of size grade of –0.01 mm ranges between 10.8 and 12.4%.

Prior to cyanidation in percolation filters, the ore samples were treated in photo-electro-chemically activated solutions. Efficiency of these solutions has been proved in the earlier studies accomplished by the researchers of the Chita Division of the Institute of Mining, Siberian Branch of RAS [5].

The solution activation procedure included: introduction of reagents to process water up to a certain concentration; stirring and saturation of the solution with atmospheric oxygen using a compressor; electrochemical and photochemical treatment of the solution. Fresh photo-electrochemically activated solution was added to the test ore samples once a day. Spent solution was discharged.

The percolation test included three stages of treatment with cyanide solutions and two stages of washing of samples with water in accordance with Zelenov's procedure [6]. The test results are given in Table 1.

The percolation tests of the proto-ore samples show that gold recovery from the initial samples with the particles -10 mm in size is 34.0%, while after cyanidation of the samples preliminary treated with photo-electro-chemically activated solutions, the average gold recovery grows by 2% and makes 36% (refer to Table 1).

For the comparison, it is noteworthy, that gold recovery in pregnant solution of samples with particles –10 mm in size in percolation tests of the solid oxidized ore of the top horizons of the deposit by Zelenov's procedure made 82% [3]. So, recovery of the useful component from the proto-ore after cyanidation in percolation filters has more than halved as against cyanidation of the oxidized ore.

Table 1. Results of percolation tests of proto-ore samples after pre-treatment in photo-electro-chemically activated solutions.

Test No.	Sample No.	Gold recovery in solution, mg	Recovery rate of gold, %
1	1048	0.442	36.8
2	1081	0.478	38.5
3	1045	0.592	49.3
4	1049	0.430	35.6
5	1025	0.277	23.5
6	1063	0.426	34.9
7	1042	0.379	31.7
8	1020	0.446	38.1
Average		0.434	36.05
Standard deviation		0.088	7.21

Average gold recovery from slime after cyanidation in bottle conditioners was 32 unit fractions (on the assumption of slime gold content of 1.15 g/t). As it is, gold recovery after agitation (for 6 hours) should be higher than after percolation filtering. Based on that, it is assumed that slime gold content is under 1.0 g/t actually. However, gold content of slime was not determined.

Heap leaching tests of agglomerated proto-ore with the particles –10 mm in size in percolation filters included three different schemes.

Scheme 1. Agglomeration of ore with strong cyanide solution. The process solution was made by feeding water in percolation filter.

Scheme 2. Prior to agglomeration, the ore sample was treated in the photo-electrochemically activated solution. Agglomeration stage also used the photo-electro-chemically activated solution. The process solution was weak cyanide solution.

Scheme 3. Pre-agglomeration treatment of the ore sample in the photo-electro-chemically activated solution. Agglomeration uses a strong cyanide solution. The process solution is made by feeding water in percolation filter.

The figure depicts the cyanidation outcome. Gold recovery rate (at the ore gold content of 1.15 g/t) is 0.37 unit fractions in scheme 1 (curve 1), 0.36 unit fractions in scheme 2 (curve 2) and 0.43 unit fractions in scheme 3 (curve 3). Each curve is plotted base on average values of two concurrent tests.

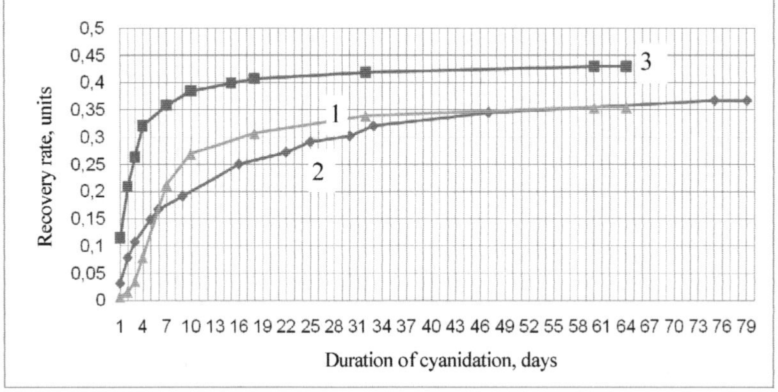

Cyanidation of agglomerated proto-ore of Pogromnoe deposit: leaching scheme 1—curve 1; leaching scheme 2—curve 2; leaching scheme 3—curve 3.

The tests have demonstrated that ore pre-treatment with the photo-electrochemically activated solutions enhanced recovery rate of gold from proto-ore and shortens duration of cyanidation.

It is seen in the figure that in reference leaching scheme 1, recovery rate of gold is 0.37 unit fractions and active passing of gold into solution (increment in recovery rate more than 1% per day) ends on the 47th day of leaching. This scheme uses no photo-electrochemically activated solutions.

In scheme 2, the photo-electro-chemically activated solution is included both in ore pre-treatment and agglomeration stages. The process solution is a weak cyanide solution. The gold recovery rate in this scheme is 0.36 unit fractions, and active passing of gold into solution comes to an end by the 32nd day of leaching.

In scheme 3, agglomeration and generation of process solutions is in accordance with scheme 1, but the ore is treated with the photo-electrochemically activated solution before agglomeration. As a result, the gold recovery rate is 0.43 unit fractions, and active passing of gold into solution makes 32 days of leaching.

Thus, pre-treatment of proto-ore by the photoelectric activated scheme (scheme 3) prior to agglomeration in strong cyanide solutions has increased gold recovery rate by 6% as compared with the reference scheme (scheme 1). Moreover, duration of the active leaching period is shortened by 15 days (see the figure).

When ore pre-treatment and agglomeration uses the photo-electrochemically activated solutions and a weak cyanide solution (scheme 2), the gold recovery rate remains nearly the same as in the reference scheme (scheme 1). However, kinetics of leaching in scheme 3 is more complicated. The duration of active leaching period is reduced (refer to the figure).

Let us compare the presented results with the tests on column heap leaching of agglomerated ore at the TOMS Institute. Agglomeration used cement, lime and a weak cyanide solution. The process solution was a weak cyanide solution, too. With the samples with particles -7 and -5 mm in size, gold recovery in solution made 41.42 and 49.3%, respectively. The active leaching period ended by the 40th day [4].

Table 2 offers the generalization of the described test data on heap leaching agglomerated ore of different size grades.

Table 2. Gold recovery in heap leaching of agglomerated proto-ore of different size grades.

Test No.	Ore size, mm	Gold recovery, %	
		Without pre-treatment	After pre-treatment by the photoelectric activation scheme
1	–10	37.0	43.0
2	–7	41.4	–
3	–5	49.4	–

Comment: agglomeration uses cyanide solutions

It follows from the analysis of Table 2 data that there is a clear relationship between the gold extraction after heap leaching and the size grade of proto-ore. Furthermore, as per the test data, ore pre-treatment by the photoelectric activation scheme enables gold extraction of 43.0% from ore with the particles –10 mm in size, which is by 1.4% higher than the extraction of the useful component in cyanidation of ore with the particles –7 mm in size without the pre-treatment (Table 2). On this basis, it has been concluded that for the material -10 mm in size, influence of the photo-electrochemical activation pre-treatment on the gold recovery rate after heap leaching of agglomerated ore is comparable with the ore crushing to the size of –7 mm.

3. Conclusion

Two experimental series to simulate concentration of rebellious proto-ore from Pogromnoe deposit by the method of heap leaching with the photo-electrochemically activated solutions have been carried out.

The first tests series used laboratory percolation filters and ore material with the size of -10 mm. The process solutions were filtered through the test ore layer in the mode of one day irrigation and one day pause. The test aimed to study the influence of the ore pre-treatment with the photo-electrochemically activated solutions on the gold recovery rate of leaching. It was found that the extraction of the useful component after the regular percolation filtration was 34%, while after the ore pre-treatment with the photoelectrically activated solution, the gold extraction grew by 2%.

The second leaching test series used laboratory percolation filters and agglomerated material with the size of -10 mm. The tests followed three schemes. The first, reference scheme used a strong cyanide solution in agglomeration. In the second scheme, the test proto-ore material was pre-treated with the photoelectrically activated solution prior to agglomeration. The agglomeration stage also used the photoelectrically activated solution and a weak cyanide process solution. In the third scheme, the ore agglomerated was preceded by the pre-treatment with the photo-electro-chemically activated solution and the agglomeration used a strong cyanide process solution.

As a result, it was found that the pre-treatment of the proto-ore sample with the particles -10 mm in size using the photo-electro-chemically activated solutions before the agglomeration in the strong cyanide solutions enhanced gold extraction by 6%.

Finally, it has experimentally been proved that the use of the photo-electro-chemically activated solutions at the stage of ore pre-treatment enables enhanced gold extraction from proto-ore of Pogromnoe deposit by 2–6% depending on the process scheme of heap leaching. It is noteworthy that the presented results are obtained on ore samples with the particles -10 mm in size. It is arguable that with smaller size grade material, the use of the photo-electrochemically activated solutions will be even more efficient as in the stages of ore pre-treatment and agglomeration, and in the stage of leaching.

References

[1] Ilina GF 2003 Pogromnoe deposit is a new morphology and genetics kind of a gold deposit in the East Transbaikalia *Prospects for Gold Mining in the Transbaikalia: International Conference Proceedings* Chita (in Russian)

[2] Spiridonov AM, Kulikova ZI, Budyak AE, Zorina LD, Pavlova LA, Parshin AV and Granina EM 2014 Material constitution of Pogromnoe gold-bearing ore (East Transbaikalia) *Geolog. Mineral.-Sur. Resurs. Sibir.* No 3s-1 pp 113–116

[3] Glotova EV, Krasnikov VI, Drobyshev VF, Kosmachev GP, Sokolovskaya VV and Bylkova IP 2003 Technological issues of commercial development of Pogromnoe gold-bearing ore deposit *Prospects for Gold Mining in the Transbaikalia: International Conference Proceedings* Chita (in Russian)

[4] *Testing of Column Heap Leaching of Pogromnoe Ore: R&D Report* Irkutsk: TOMS NIIPI 2011

[5] Sekisov AG, Shevchenko YuS and Lavrov AYu 2016 Prospects for Underground Leaching in Gold Mines *Journal of Mining Science* Vol 52 No 1 pp 115–120

[6] Zelenov VI 1989 *Gold-Bearing Ore Testing Procedure* Moscow: Nedra (in Russian)

Role of hydrogen ions in standard and activation heap leaching of gold

YuI Rubtsov

Chita Division, Chinakal Institute of Mining, Siberian Branch, Russian Academy of Sciences, Chita, Russia

E-mail: rubtsov.jri@yandex.ru

Abstract. The role of hydrogen ions in activation heap leaching of gold from rebellious ore has been studied, which has allowed enhancing gold recovery. The author puts forward a gold leaching circuit with the use of activated oxygen-saturated solutions acidified to pH = 6–9.

One of the causes of reduction in old extraction by cyanide heap leaching is the absence of theoretical evidence on mechanisms of intracrystalline processes.

Hydrogen ions in monomolecular surface layer of crystalline lattice possess comparatively high energy of bond with mineral atoms (of the order of 40–120 kJ/mole) and, being very small in size, can actively participate in intracrystalline processes. In this respect, oxidizing properties of hydrogen ions can prevail over oxidizing properties of oxygen. The introduced hydrogen ions have an effect on atoms (ions) in deep layers of crystalline lattice either by means of sequential transmission of impulses between internal layers or by way of diffusion inward the lattice, through interstitial sites, up to reaching sites of internal layers [1]. In activation cyanide leaching of rebellious gold from ore, an important part belongs to water as the main source of hydrogen ions. A one-time source of hydrogen may be acid; at the same time, after equilibrium exchange processes of electrolytic dissociation and hydrolysis, it is difficult to stabilize concentration of hydrogen ions)(pH) without secondary electrochemical treatment of cycling solution in anode chamber of diaphragm cell. The secondary intercrystalline processes conditioned by interaction between hydrogen ions, cyanide ions, hydroxyl radicals and lattice atoms may induce hypermobility of the lattice components, including gold. When hydrogen ions come to the surface of an electroconductive component of ore (iron sulfide crystal) and accept electrons from gold atoms, the generated gold ions in the lattice become more mobile or active as a consequence of charging and size reduction. In case that disjoining action of electrolytic solution enables cyanide ions to flow in micro-channels or structural damage and to reach gold surface, the nascent god cyanide complexes $[Au(CN)_2]^-$, structurally linear and extremely stable, are forced by diffusion directed to the periphery of crystal to leave the crystal and pass into solution. In the studies performed by the present paper author, water pH was adjusted by acidification of solution and aqueous medium was activated by electrolysis.

These studies were aimed at enhancement of gold extraction by rapid heap leaching from low-sulfide ore using activated oxygen-saturated solutions with hydrogen ions involved.

The bottom line of the studies is the new approach to gold cyanidation, when classical cyanidation is likely to be accompanied with the other chemical reactions, e.g., with the participation of hydrogen ions [2]

$$Au+2CN^- +H_2O +1/2O_2 = [Au(CN)^-_2]+2OH^- (\Delta G^o{,}_{298}{}^oK = -151.35 \text{ kJ/mole}),(1)$$

$$Au+2CN^-+2H^+ = H[Au(CN)^-_2]+ 1/2H_2(\Delta G{,}_{298}{}^o{}_{K,react} = -73.75 \text{ kJ/mole}),(2)$$

$$Au+2CN^{-}+2H^{+}+1/4O_2 = H[Au(CN)^{-}_{2}]+1/2H_2O(\Delta G_{,298}{}^{o}{}_{K,react} = -192.37\ kJ/mole).(3)$$

In this case, hydrogen ions initiate gold cyanidation in solution: gold leaching becomes feasible without oxygen (reaction 2), while when oxygen is present, more expectable is gold cyanidation with the release of water rather than hydroxide ion (reaction 3, $\Delta G_{297}{}^{o}{}_{K,react} = -192.37$ kJ/mole).

In accordance with Figure 1, at pH = 9.35 concentration of C_{CN^-}-ions is comparable with C_{HCN} [3]. When pH ≤ 9.35 concentration of $C_{H}{}^{+}$-ions grows and gold cyanidation rate should increase; when pH > 12, cyanidation of gold is terminated.

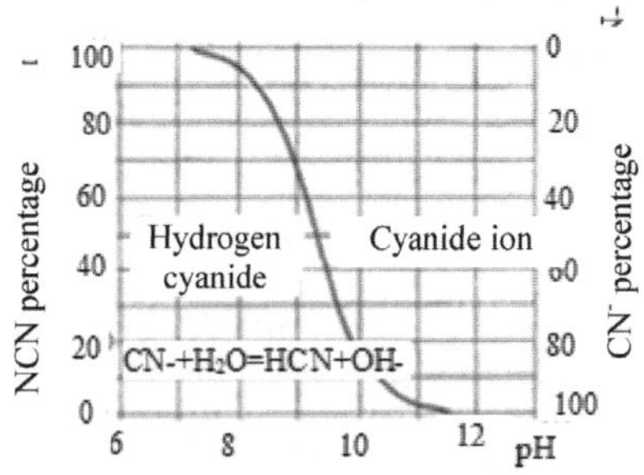

Figure 1. Influence of pH on percentage of HCN and CN⁻ in aqueous solution.

Table 1 gives data on gold cyanidation with hydrogen ions and without oxygen. Based on these data, it has experimentally been proved that gold and hydrogen ions interact according to reaction 2 as the residual quantity of dissolved oxygen after high-pure argon blow of the solution is 10 times less than it should be in accordance with stoichiometry of cyanidation reaction 1 as indicated in the findings of the Hydrology Department Laboratory at the Transbaikalia State University. Incidentally, on the surface of gold chips or gold webs, gas bubbles 1–2 mm in size were visually observed. The calculations showed that the gas could be hydrogen as its mole fraction conformed with the mole fraction of gold in dissolution according to reaction 2 and made 0.2.10–4 moles.

Table 1. Influence of pH of aqueous cyanide solution of gold cyanidation in argon medium.

No.	C_{NaCN}, g/l	Introduced HCN, ml	Introduced H_2SO_4, g/l	*T*, days	pH	Gold weight, mg	After-test gold weight, mg	Comments
1	1	–	–	3	11	553.8	553.8	Gold slab, *S* = 1 cm^2, *t* = 26°C,
2	1	–	–	3	11	11	11	Gold web, *t* = 25°C, argon
3	1	–	–	3	11	7	7	Gold chips, particle size < 0.02 mm, *t* = 24°C, argon
4	1	–	2	3	6	10.4	6.5	Gold web, *t* = 22°C, argon
5	–	2	–	3	7	10.2	8.1	Gold chips, particle size < 0.02 mm, *t* = 17°C, argon
6	–	2	–	4	7	9.1	5.3	Gold web, *t* = 17°C, *t* = 17°C, argon
7	–	2	–	3	7	10.2	9.5	Gold web, *t* = 17°C, air

Table 2. Initial conditions of ore agglomeration for subsequent standard-circuit gold leaching.

Q of dehumidified ore, kg	Fragmentation size, mm	Absorption of NaCN, kg/t	Cement consumed in agglomeration, kg/t	Lime consumed in agglomeration, kg/t	Agglomeration solution, l	C_{NaCN} in agglomeration, g/l	*V*day solution consumed for spraying, l	Ore gold content, mg
46.32	10	0.07	0.12	0.06	0.75	0.35	6.8	69.48

Experimental leaching of gold was carried out on crushed rebellious low-sulfide ore extracted from an open pit depth of 150 m in accordance with the standard leaching circuit and with activation of leaching solution at pH = 6–9. The tests used percolation filters with the following dimensions: $h = 1.15$ m; $d_{pf} = 0.208$ mm, $S_{pf} = 0.03396$ m^2. The calculated initial parameters for ore agglomeration and standard gold leaching circuit are given in Table 2.

The standard gold leaching results are reported in Table 3.

Table 3. Results of drop gold leaching in percolation filter at pH = 10.5.

T, days	V_{day} of pregnant solution, l	pH of pregnant solution	CAu in pregnant solution, mg/l	C_{NaCN} in pregnant solution, mg/l	ΔqAu days, mg	∑qAu after percolation, mg	X Au leaching rate, %	Extra cyanide input, g	NaCN in cycling solution, g
1	5.45	12	1.31	0.15	7.14	7.1	10.2	2.1	0.81
2	6.38	8	0.64	0.15	4.08	11.2	16.1	2.1	0.95
3	6.8	8	0.64	0.15	4.35	15.6	22.4	2.1	1.02
4	6.2	8	0.37	0.15	2.29	17.7	25.7	2.1	0.93
5	6.9	9	0.36	0.15	2.48	20.4	29.3	2.1	1.03
6	6.9	7	0.34	0.15	2.35	22.7	32.7	2.1	1.03
7	6.56	8	0.22	0.15	1.44	24.1	34.8	2.1	0.98
8	6.6	9	0.16	0.15	1.06	25.2	36.3	2.1	0.99
9	6.5	6	0.23	0.15	1.50	26.7	38.4	2.1	0.97
10	6.75	6	0.2	0.15	1.35	28.0	40.4	2.1	1.01
11	6.5	6	0.16	0.15	1.04	29.1	41.9	2.1	0.97
12	7.1	6	0.16	0.15	1.14	30.2	43.5	2.1	1.06

Table 4. Initial conditions for ore agglomeration and subsequent activation leaching of gold.

Q of dehumidified ore, kg	Fragmentation size, mm	Cement consumed in agglomeration, kg/t	Lime consumed in agglomeration, kg/t	Agglomeration solution*, l	C_{NaCN} in agglomeration, g/l	NaCN consumed in agglomeration, g/	V of oxygen-saturated solution for spraying, l	Ore gold content, mg
47.29	10	0.12	0.06	0.75	13.5	10.125	4	70.9

*) agglomeration used cyanide solution after electrolytic activation and subsequent saturation with oxygen up to dissolved oxygen content of 33–39 mg/l

Table 5. Results of experimental activation gold leaching.

T, days	V_{day} of pregnant solution, l	pH of pregnant solution NaCN	CAu in pregnant solution, mg/l	CNaCN in pregnant solution, g/l	**Δ**qAu day, mg	∑**Δ**qAu after percolation, mg	X Au leaching rate,
3	3.24	10	1.92	9.5	6.2		8.7
4	2.96	8	1.98	0.22	5.9	12.1	17.0
5	2.72	8	1.87	0.15	5.1	17.1	24.2
6	3.1	9	1.72	0.15	5.4	22.5	31.7
7	2.52	7	1.71	0.15	4.4	26.8	37.7
8	2.6	8	1.7	0.15	4.4	31.2	44.0
9	2.72	9	1.65	0.28	4.5	35.7	50.3
10	2.15	6	1.61	0.2	3.4	39.1	55.2
11	1.9	6	1.62	0.22	3.1	42.2	59.5
12	1.7	6	1.53	0.23	2.6	44.8	63.2

Tables 4 and 5 give the calculated initial parameters for ore agglomeration and activation leaching of gold at pH = 6–9 and the experimental results of the activation gold leaching, respectively.

The experimental gold leaching results are given in graphical form in Figure 2. After having generalized data on standard and activation leaching of gold at pH ranged from 6 to 9, it is possible to draw a conclusion on the efficiency of activation aqueous cyanide solutions by electrolysis and subsequent saturation with oxygen and acidification up to pH = 6–9 in enhancement of gold extraction and in improvement of rapid heap leaching of gold from rebellious ores.

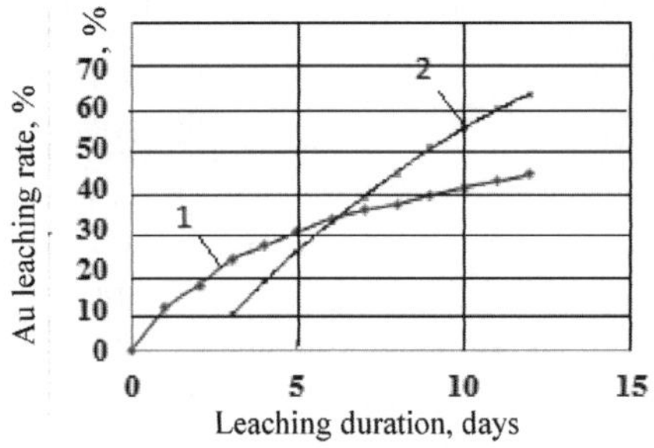

Figure 2. Gold leaching in percolation filter 1 at pH = 10.5 (curve 1) and in percolation filter 2 at pH = 6–9 (curve 2).

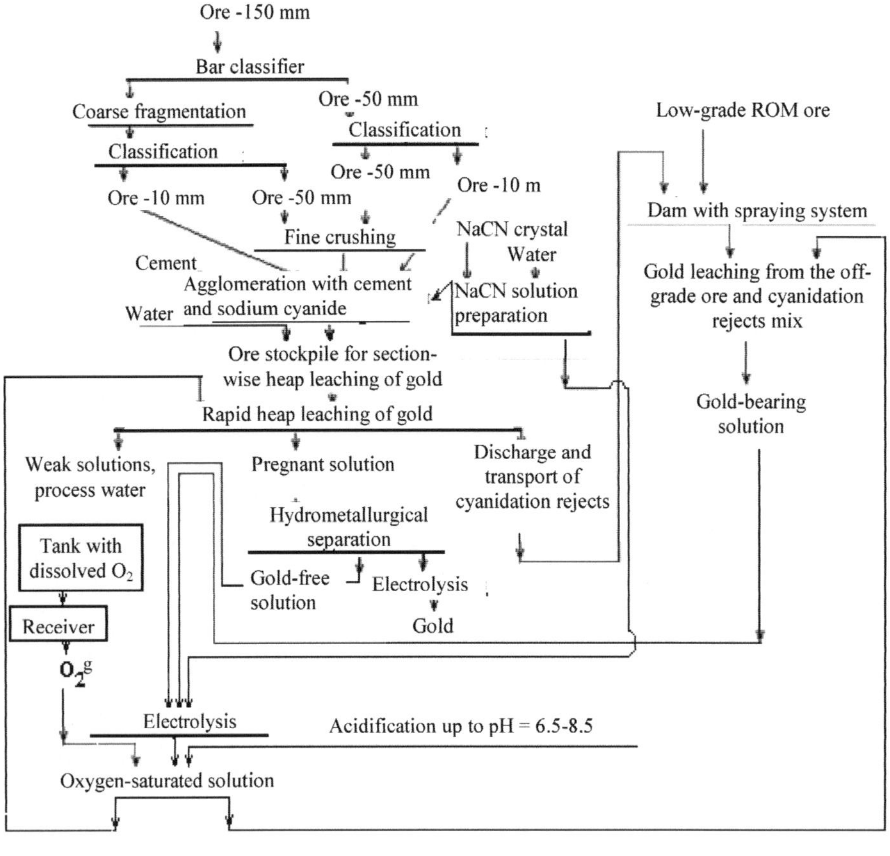

Figure 3. Flow chart of activation leaching of gold.

Based on the data from Tables 3 and 5 and using the equation of kinetics of heap gold leaching [4], the rate of the activation leaching of gold in 20 and 30 days makes 70 and 76%, respectively. In case of the standard circuit of drop heap leaching of gold, the same rate of leaching is likely to be expected in 60 and 90 days.

Then, experiments on gold leaching in cyanide solutions saturated with oxygen and acidified up to pH = 6–9, but without electrolytic activation were carried out. Efficiency of rebellious gold leaching was much higher than in the standard cyanidation circuits but lower by 7–12% than in the circuits with the activated cyanide solutions.

The introduction of the activated solutions after saturation with oxygen and acidification up to pH = 6–9 complicated the rapid heap gold leaching flow chart added with the stage of electrolytic activation of cyanide solution and the stage of acidification of cyanide solution up to (see Figure 3).

Conclusion

The introduction of hydrogen ions in activated and oxygen-saturated solutions with intent the reach the solution pH = 6–9 has a beneficial effect on gold extraction from rebellious low-sulfide ores.

References

[1] Sekisov AG, Zykov NV, Korolev VS 2012 *Dispersed Gold* Moscow: Gornaya Kniga (in Russian)
[2] Rubtsov YuI 2010 Role of hydrogen ions in gold cyanidation reactions *GIAB* No 7
[3] Ovseichyk VA, Reznik YuN and Myazin VP 2005 *Uranium and Gold-Bearing Ore Mining and Processing Technologies: Educational Aid* Chita: ChitGU (in Russian)
[4] Rubstov YuI and Reznik YuN 2007 Basic equation of kinetics of rapid heap gold leaching from oxidized quartzite ores *GIAB* No 5

IOP Conf. Series: Earth and Environmental Science **53** (2017) 012034 doi:10.1088/1755-1315/53/1/012034

Cluster chemical reactions at mineral–liquid interface in metal leaching by photo-electroactive water-and-gas emulsions

AG Sekisov

Chita Division, Chinakal Institute of Mining, Siberian Branch, Russian Academy of Sciences, Chita, Russia

E-mail: sekisovag@mail.ru

Abstract. Possibility of cluster (inter-cluster) reactions at the interface of mineral and liquid phases in leaching of metals mainly in dispersed cluster form by photo-electrically activated water-and-gas emulsions is theoretically evaluated. The governing role of active clusters of water and clustered hydrate envelopes generated under dissolution of active oxygen forms is determined. The scope of the study covers possible processes of transformation of clustered gold in mineral substance under direct interaction with the components of the active water-and-gas emulsions.

According to findings of many Russian and foreign researchers in geology and geochemistry, gold occurs in gold-bearing minerals in the form of micron and submicron particles and in dispersion form of poly- and mono-element clusters. Dispersed gold particles are geochemically associated in gold-bearing minerals with such metals as silver, copper, iron, lead, bismuth, and with anion formers such as sulfur, silicium, arsenic, antimony and tellurium. Metals and gold particles are in metallic bonding, while chemical bond of gold and anion formers is much more complex and is at the early stage of being explained theoretically.

Inasmuch as, first, it is known on the existence of such natural minerals as petrovskaite (AuAgS), utenbogaardtite (Ag_3AuS_2), calaverite ($AuTe_2$), aurostibnite ($AuSb_2$) and montbreite (($(AuSb_2)Te_3$)), and, second, it is experimentally been found to be feasible to obtain cluster compounds of gold with sulfur (Au_nS_m) and with sulfur and hydrogen, it can be supposed that sulfide mineral can contain poly-element cluster nanostructures with a mono-element "core" with the gold–gold bond and a "shell" composed of atoms either of basic mineral-forming elements (iron, copper, sulfur, arsenic) or atoms of antimony, tellurium, silver, hydrogen, i.e. doping elements that indirectly connect gold with basic atoms (ions) of lattices of the host mineral.

In view of high degree of geochemical bonding between silicium and gold, the evidence of clustering of silicium with alkali metals (fullerene-like clusters of silicium and encapsulated atoms of alkali metals: NanSim, KnSim, RbnSim, CsnSim), which are to a certain degree the analogs of gold, it can be supposed that metal minerals can contain gold–quartz and gold–sulfide–quartz assemblages, and gold–silicium bi-element clusters (AunSim). Similar gold–carbon nanostructures (AunCm) could appear in the course of bio-geo-chemical recycling of gold (upon disintegration of organometallic complexes in zones of orogenic activity, a part of gold migrates in solid-phase medium, and another part strengthens bonding with carbon). So, hydroxyl radicals and sodium and hydrogen ions can dissociate dispersed gold clusters, i.e. break both excitonic bonds between gold atoms and, for instance, silicium atoms, and metallic gold–gold bonds in the core of a cluster. As the atoms of "shell"-making elements are also clustered, they, when in contact with aqueous medium and, consequently, in the interaction with outside hydrated complexing agents, which are actually cluster structures, can form "above-cluster" compounds with the "core" of gold atoms. It is assumable that clusters contain "collective-state electrons" according to Little, which conditions chemical interaction between them as a single particle and another object of similar structure. In this manner, the first stage of dissolution of cluster gold may be single-step interaction between a complex of atoms, making the "shell" of the gold cluster, and hydrated clusters of complexing agents of solution followed by indirect passing into film water in the form of an aggregate hydrated complex with the preserved bonding of the "core" of gold atoms and the "shell" atoms. Naturally, probability of this process depends on energy and structural conformity of hydrated complexes making the "shell" of the gold clusters. Therefore, these hydrated complexes, containing molecules of water and ions and/or radicals of certain elements or their compounds interacting with the atoms in the gold cluster "shell," should have a cluster structure, as well. Clustering of water molecules can take place during formation of hydrated shells around molecules (atoms, ions) of soluble solids, clathrates of molecules (atoms) of soluble gases, or under interaction with inoculating ions of hydrogen and oxygen, e.g. H+, OH- [1, 2], as well as with radicals OH* or ion radicals, e.g. high-active oxygen ion radical (O*-). Proton and hydroxyl ions of neighbor molecules can make metastable bonds (H+H- or H*(e+e-)H*), which results in clustering. Energy of a cluster, e.g. H+(H2O)n, OH*(H2O)n (where n = 2–30), and, accordingly, its reactivity depend both on the energy of the inoculating center and the energy of water molecules and their dissociates (H+, OH-, OH*).

Such ionized and radical reactive components of water clusters can be obtained using different methods of treatment of aqueous solution of initial reagents, though the most suitable methods are electrolysis, photolysis and photoelectrolysis. Electrolysis of solutions of some salts and acids greatly intensifies dissociation of water molecules, which increases energy of hydrated shells of ions in water, and can produce compounds with polymeric structure and/or with high concentration of oxygen, which can result in formation of high-energy water clusters connected with these compounds.

In particular, electrolysis of sulfuric acid (H_2SO_4) produces in the anode zone metasulfuric and persulfuric acids (H_2SO_5, $H_2S_2O_8$); azotic acid—pernitric acid (HNO_4); sodium bicarbonate ($NaHCO_3$)—metacarbonic and percarbonic acids (H_2CO_4,$H_2C_2O_6$) in the anode zone and superoxides of alkali metals ($Na_2C_2O_6$, $K_2C_2O_6$) in the interelectrode space. Electrolysis of sodium chloride solution produces hypochloric acid (HClO) perhypochloric acid ($H_2Cl_4O_4$) in the anode zone and sodium hyperchlorine (NaClO) in the interelectrode space. UV irradiation of water-and-gas emulsion after the electrolysis generates such highly active oxidizers in the gas bubbles as atomic oxygen, superoxide radical ion (O^{*-}), ozone and hydroxide radical (OH*), which, under interaction with film water around the gas bubbles, create ion radical and radical-bearing clusters.

When such hydrated cluster appeared after photo-electrochemical treatment of an initial solution contains a few complexing radicals and/or ions, it can be a reaction entity with the structural and energy parameters complementary to the structure and energy of the "shell" of a gold-bearing cluster.

The active clusters and clathrates–clusters in treated solutions can support passing of gold-bearing cluster in water phase both directly and indirectly as is described below.

Periodic dissociation of water molecules making an active cluster (or cluster shell of clathrate) may generate secondary active particles: H^+ and OH^-, OH^* "fixed" ionic pares of H^+H^- type, active (excited) molecules and/or ions: H_2O^*, H_2O^+, H_2O^-, which can diffuse inside lattices of dispersed gold-containing minerals and induce breaking of bonds between atoms (ions) of mineral-forming elements and nano-cracks in the lattices. Moreover, active ions and radicals of hydrogen and oxygen and their compound can join together inside a crystal of a metallic mineral and generate meta-stable reactive complexes of the type of $nH_2O^*H^+H^-$, supporting clustering of atoms of such elements and silicium, sulfur, oxygen, iron, etc., making a 'shell" around the clustered atoms of gold. The "shell" elements make bonds with water oxygen and break bonds with the "core" gold atoms. Sodium and/or potassium ions diffusing into the double electric layer at the contact with the surface of mineral particle can liberate from the hydrated shell owing to oxidation of negatively charged hydroxyl ions in the shell by protons and can also enter lattices. Sodium cation, being high active and having smaller ionic radius, can replace gold in silicium clusters. The similar effect can be cerated by hydrogen ions (protons) as they also can make clusters with silicium of the type of [3].

Later on, ionic pairs $H^+ \cdot H^-$ and hydroxyl radicals, diffusing through the double electric layer to the lattice, excite sharing electron shells (including hybrid, at levels 5d-6s) of clustered gold atoms. This results in the exponentially growing probability of break of the chemical bonds between gold atoms and surrounding atoms of mineral-forming elements. The liberated mono-elements clusters of gold and/or individual gold atoms are permitted to migrate inside the lattice of a mineral and reach micro-cracks and pores, including in the form of meta-stable compounds with hydroxide radical. It is known that even natural water contains stable occurrence form of gold—$AuOH(H_2O)^0$—in a wide range Eh–pH. So, during interaction of hydroxyl radicals and protons, which penetrate a metallic mineral lattice from process solution, with the dispersed gold atoms (including clustered), it is highly possible that mobile meta-stable compounds AuOH* and AuH* will appear and can be moved to contact surface under heap leaching. Consequently, this ensures further complexing of gold and cyanide in leaching solution and diffusion of gold-and-cyanide complexes to pregnant solution.

To make secondary gold clusters at the surface of micro-cracks and pores and the "cores" of the primary poly-element gold-bearing clusters (after its separation from the "shell") pass into solution, it is most efficient to use not individual hydrated anions $nH_2O{*}CN^-$ or radicals $nH_2O{*}CN{*}$ but structurally and energy complimentary hydrated clusters including a few single compounds CN as complexing entities: $m[(CN^*)(OH^-H^+)nH_2O^*(Na^+)OH^-]$.

Ionic pairs $H^+ \cdot H^-$ in the composition of these reactive clusters transform ions CN^- into active meta-stable radicals CN^* and, at the same time, excite shearing electron shells of gold clusters and, consequently, open way for single-stage interaction and clustering of hydrated anions $[nAu{*}m(CN)_2]^{n-}\ nH_2O$

$$nAu{+}2m\,[OH{*}nH_2O^*(Na^+)OH^-H^+(CN^-)] = nAu{+}2m\,[(CN^*)(OH^-H^+)nH_2O^*(Na^+)OH^-] =$$
$$= [nAu{*}m(CN)_2]^{n-}{*}2m[(n{+}1)H_2O] + 2m\,NaOH{*}\,. \qquad (1)$$

The known models of gold cyanidation by Elsner (2) and Bodlander (3), as follows from the formulas below, also assume interaction of many sodium cyanide molecules in complexing but fail to expose physicochemical background of these reactions

4 Au + 8NaCN + O2 + 2 H2O = 4NaAu(CN)2 + 4NaOH , (2)

2Au + 4NaCN + O2 + 2H2O = 2NaAu(CN)2 + 2NaOH + H2O2 .

H2O2 + 2Au + 4NaCN = 2NaAu(CN)2 + 2NaOH,

4Au + 8NaCN+ O2+= 2H20 = 4NaAu(CN)2 + 4NaOH . (3)

The appropriateness of the proposed model (1), assuming dispersed gold leaching by photoelectrically activated solutions as the result of the cluster reaction with the active hydrated complexing agents, is aproved both by the experimental evidence of clustering between water molecules and their dissociatioin products [3] and by the existence of a narrow range pH = 10–11.5 for efficient reaction of cyanidation. The latter fact implies a significant role of the joint presence of both ions H^+ (protons) and OH^- (and, consequently, meta-stable ions H^-) in the hydrated clusters, including complexing agents for gold, with the periodic generation of meta-stable pairs ensuring their reactivity.

Numerous experiments on photoelectric activation leaching of dispersed gold from rebellious metalic minerals and waste carried out at the Chita Division of the Institute of Mining, SB RAS, have proved the theoretical statements of the discussed model of cluster reactions. It has been found that leaching rate and extraction rate of gold depend on gold dissociation intensity and re-combination of water molecules, on mobility of hydroxyl ions and hydroxonium, on concentration of soidum cyanide and on its dissociation in photoelectrically activated solution. The most interesting proof of formation of cyanide-bearing clusters in photoelectrically activated solutions is the essential decrease in concentration of "free" cyanide ions, which is determined by alkalimetric testing, in the first hours after preparation of the activated soution (water and gas emulsion) with the subsequent increase up to initial values that are close to estimated values. The same increase in the concentration of sodium cyanate, an oxidized form of sodium cyanide, is not observed. The main experimental finding to prove cluster behavior of chemical reactions during leaching of gold and associated silver is the stage-wise yield of these metals in pregnant solutions as against gradully descending kinetic curve in the test experiments on plain water solutions with equl concentrations of cyanides.

References

[1] Korolev VA 1996 Associated water in rocks: New evidences and issues *Soros. Obraz. Zh.* No 9
[2] Tantsyrev GD and Nikolaev EN 1971 Formation of clusters in films of frozen polar substance under bombarding with ions *Pisma ZhETF* Vol 13
[3] Meleshko VP, Morozov YuN, Shveigerd VA 1999 Water–silicium clusters *Zh. Struktur. Khim.* Vol 40
[4] Sekisov AG, Zykov NV and Korolev VS 2012 *Dispersed Gold. Geological and Technological Aspects* Moscow: Gornaya Kniga (in Russian)
[5] Sekisov A, Paronya A, Kuzin V and Lalbekyan N *Patent 5.942.098 USA International Class C 25 B 001/00, C 25 C 001/20 Method of Treatment of Water and Method and Composition for Recovery of Precious Metal* Filed 12.04.96
[6] Norov YuV 2010 Quantochemical analysis of structure and electron composition of small nano-clusters MoxO3x-y ($x = 1–6$, $y = 0, 1, 2$) and AunSH, *PhD in Chemistry Dissertation* Moscow: MGU (in Russian)

Small-scale and large-scale testing of photo-electrochemically activated leaching technology in Aprelkovo and Delmachik Mines

AG Sekisov*, AYu Lavrov and YuI Rubtsov

Chita Division, Chinakal Institute of Mining, Siberian Branch, Russian Academy of Sciences, Chita, Russia

E-mail: *sekisovag@mail.ru

Abstract. The paper gives a description of tests and trials of the technology of heap gold leaching from rebellious ore in Aprelkovo and Delmachik Mines. Efficiency of leaching flowsheets with the stage-wise use of activated solutions of different reagents, including active forms of oxygen, is evaluated. Carbonate–peroxide solutions are used at the first stage of leaching to oxidize sulfide and sulfide–arsenide ore minerals to recover iron and copper from them. The second stage leaching uses active cyanide solutions to leach encapsulated and disperse gold and silver.

Success gained in commercial testing and introduction of innovative mining and processing technologies for rebellious ore, based on brand new physical, physicochemical, biochemical and combination processes results in no enhancement in recovery of valuable components from low-grade and impoverished ore and mining and processing waste due to high capital and operating cost involved.

In this respect, it remains of concern to develop, evaluate and substantiate innovative processes to be included in flow charts of different physicochemical technologies, namely, underground (mine and borehole), heap and cuvette leaching. These processes should involve activation of leaching solution, minerals and sorbents by environmentally safe and low power-hungry physical fields and emissions. Combination photochemical and electrochemical treatment of solutions of some reagents allows producing highly active oxidizers at minimum energy input.

The most representative experiments on photo-and-electrochemically activated heap leaching of agglomerated and unagglomerated ore from waste stockpiles at Pogromnoe and Delmachik deposits, with high residual gold content (0.72–0.78 and 0.8–1.1 g/t, respectively), were carried out in Geotechnology Laboratory of the Chita Division of the Institute of Mining, SB RAS, and directly in Aprelkovo and Delmachik Mines. The laboratory tests showed that additional recovery of gold from Aprelkovo Mine waste ore by cyanide solutions prepared based on photo-and-electrochemically activated solutions of sodium bicarbonate with different admixtures, which was actually treatment by water-and-gas emulsions, made 37.3 % without secondary agglomeration and 58.3 % after secondary agglomeration. The concurrent tests of the same material in large percolation filters with the standard cyanide solutions yielded additional gold recovery of 20.1 % without secondary agglomeration and 35.6 % with the secondary agglomeration. After similar experimentations using Delmachik Mine waste ore material, additional gold recovery made 12–17 % as against the standard cyanidation circuits. It is noteworthy that the check solutions featured high concentration of cyanides (0.4 %), whereas the testing solutions had cyanide content not higher than 0.03 %. On the joint technical meeting at Aprelkovo Mine, it was decided to use the circuit of activation leaching without secondary agglomeration though it produced much lower gold recovery but its introduction in the process flow chart would require re-excavation of the waste material stockpile. In this regard, the authors suggested minimized drop-like feed of photo-and-electrochemically activated solution at the first treatment stage, which would be sufficient to soak the material, with the stage of ageing for 2–3 days before spraying the material with the high-concentration cyanide solution with pH = 10.5–10.7. For the tests in the mine, two samples were taken from different waste material stockpiles of unagglomerated and agglomerated ore (with the residual gold content of 0.65 and 0.55 g/t, respectively). Each sample was divided into one check and two testing specimens. As follows from Figure 1, the highest recovery was reached with the photo-and-electrochemical activation despite lower content of cyanide in the activation solution as compared with the check solution. In case of the waste material with the higher gold content, additional gold recovery with the activation circuit was 29.0 and 23.3 % in the testing specimens and (curves 3 and 2) and 21.9 % (curve 1) in the check specimen; as for the lower grade sample, additional gold recovery was 19.8 and 20.2 (curves 5 and 6) in the testing specimens and 14.3 % (curve 4) in the check specimen.

Gold yield in the testing pregnant solution takes 2 cycles as against gradual extraction of gold in the check circuit (see Figure 2). This means that activated solutions extract dispersed gold.

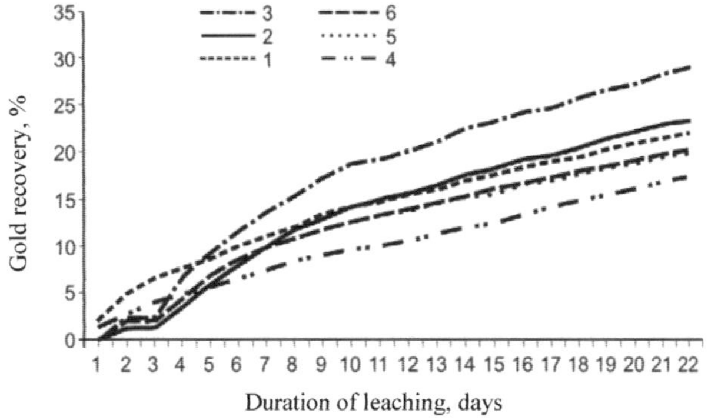

Figure 1. Gold recovery from materials of heap leaching waste stockpiles (3 upper curves for the specimen with the gold content of 0.65 g/t; 3 lower curves for the specimen with the gold content of 0.55 g/t). The check curves for the specimens are the first and the fourth relative to the abscissa axis, respectively.

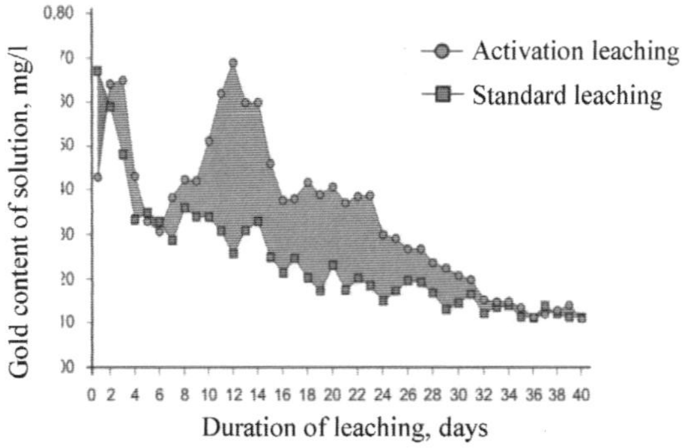

Figure 2. Gold content of pregnant solution versus time in the standard circuit and activation leaching.

Au leaching from rebellious Delmachik ore using the standard agglomeration circuit ended with the low gold recovery in liquid phase despite the high content of cyanides, namely, 27 % as against the expected recovery of 32–35 %. This could be explained by the increased concentration of copper and antimony in the test ore material, and by solubility of these elements in the high-concentration cyanide solution. The introduction of the photo-and-electrochemical activation with the agglomeration in the circuit of the same sample produced gold recovery up to 57 %. Based on the testing results, management of Aprelkovo and Delmachik Mines decided on commercial testing of photo-and-electrochemical activation in processing of ore from Antimony Site using the flow chart of flotation (with mobile cell) and cyanidation of the flotation concentrate with photo-and-electrochemical activation of input and recycling solutions.

For the commercial tests in Aprelkovo Mine, the present paper authors and the engineers of NordGold company designed a process-and-equipment flow chart of the photo-and-electrochemically activated heap leaching of low-grade oxidized ore and waste stockpiled material. Additional pumps were purchased, and electrochemical and photochemical reactors to process initial solution of sodium bicarbonate were assembled (Figure 3).

Figure 3. Electrochemical and photochemical reactors for activation of process solutions at Aprelkovo Mine.

The sequence and parameters of feed of the process solution on a stockpile composed of leached and low-grade ore are developed.

Spraying is executed by unit section (a unit section is 30 000 thousand tons of ore, total area of 2206 m^2). Parameters: height—8 mm, bulk density—1.7 t/m^3, pumping capacity—700 m^3 per day, maximum capacity of 2 main lines—760 m^3/day (reference data), spraying density $700\times1000/2206 = 317$ l/m^2 per day. Capacity of grinding and sorting plant—170 thousand tons.

Commercial testing of process solutions included:

—Stage 1—three-days long feed of photo-and-electrochemically activated peroxide–carbonate solution (without sodium cyanide). Initial moisture content of ore in winter–3 %; moisture content or ore treated with the process solution is 9.7 %; i.e. it is required to feed 2100 m^3 of the photo-and-electrochemically activated solution per unit section;

—Stage 2—the solution is added with sodium cyanide up to the concentration of 0.35 g/l. Spraying is carried out for 8 days. Cyanide consumption makes $700\times8\times0.35 = 1960$ kg.

By the time of this paper preparation, the increment in gold content of the pregnant solutions per the test unit sections was not less than 10 %.

References

[1] Lavrov AYu 2015 Acceleration of dispersed gold leaching by photoelectrochemical activation of process solutions *Gorny Zhurnal* No 8 pp 32–36

[2] Sekisov AG, Shevchenko YuS and Lavrov AYu 2016 Prospects for underground leaching in gold mines *Journal of Mining Science* Vol 52 No 1, pp 115–120

[3] Sekisov AG, Lankov BYu, Grinchenko IV, Lavrov AYu et al 2016 RF Patent No. 2580356. Method of heap leaching of gold from rebellious ores *Bulleten Izobret.* No 18

Explosion-assisted preparation of dispersed gold-bearing different-grade ore for selective mining

AI Trubachev[1,2] and NV Zykov[3]

[1] Transbaikal State University, Chita, Russia
[2]Chita Division of the Chinakal Institute of Mining, Siberian Branch, Russian Academy of Sciences, Novosibirsk, Russia
[3]Agoshkov Transbaikal Mining College, Chita, Russia

E-mail: [1]zabgc@mail.ru

Abstract. It is found that there are transient zones (between quality and off-quality ore areas) with the respective content of useful component in an ore body, and a variant of explosive treatment of such zones before the selective mining is put forward. Practicability of two processing technologies is evaluated: processing of high-grade and low-grade ore from the transient zones and heap leaching of metals from the low-grade and impoverished ore. Open mining technology is conventional truck-and-shovel scheme, with distributed ore flows to processing plant and (or) to heap leaching, which generally enhances the mine efficiency.

In open-pit mining of gold-bearing ore, owing to complicated morphology of ore bodies, great variety of forms of gold occurrence in them and stochastic spatial distribution of useful components, blasting and extraction result in considerable dilution and intermixing of various grade ore particles. This particularly relates nonlinear interfaces of ore and rock mass as well as complex-geometry interfaces of different grade and different type ores. The application of such physicochemical geotechnologies as heap leaching to ore mining allows profitable processing of diluted and off-grade ore, which enhances efficiency of overall mining [1]. When a mine uses concurrently two processing technologies, namely, concentration of high-grade and ROM ore and heap leaching of low-grade and diluted ore, it is advisable to use selective extraction at different type ore interfaces: the first kind ore mass is not intermixed with enclosing rocks or another type ore under blasting; the second kind ore mass is purposely mixed by blasting with off-grade ore and enclosing rocks, with large yield of fine particles and many microcracks and pores induced in blasted fragments. The first kind ore mass is forwarded directly to a processing plant (or to hydrometallurgical processing after crushing and milling); the second kind ore mass is subject to heap leaching in view of low content of valuable component. Prior to heap leaching, such ore may go to crushing and separation, with production of standard quality concentrate, middlings and tailings, and to agglomeration of fines. For implementation of the proposed approach to ore quality control at interfaces of different ore zones at the stage of blasting and selective extraction, it is required to carry out detailed and operational exploration and use these data to distinguish at least three zones in ore body and host rock mass: standard quality ore (suitable for conventional processing), off-grade ore or enclosing rocks and transition zones. In case of identifying transition zones between quality and off-grade ores or enclosing rocks, two scenarios are possible. When the identified zones are in direct contact (low valuable component gradient), their interface is marked after blasting with pennants, color ribbons or markers along the middle line between mouths of blast holes with allowance for rock mass deformation under explosion. In case that such contact is of low probability (rich ore—barren rock or rich ore—off-grade ore), a transition zone is anticipated: additional blasting is carried out in this zone with the central shift off the rich ore boundary to the off-grade or barren rocks, then short-delay blasting is executed so that to ensure intensive fragmentation and mixing of mineral and rock mass; after that, selective extraction is performed: unshattered and undiluted ore mass at the contact with the rich ore is extracted together with this quality ore, and the mixture of enclosing rocks and off-grade ore after blasting is sent to heap leaching [2, 3].

Sometimes it is possible to send ore mass from a transition zone to separation in order to extract fragments with standard and off-standard content of valuable components. The identification of the transition zones and the differentiated blasting of ore and barren rock areas will enable elimination of dilution of high quality rich ore with barren rocks or off-grade ore and, thus, will allow improvement of recovery in the main processing circuit. At the same time, metal from ore mixed with off-grade and barren rocks under blasting is extracted by heap (or cuvette) leaching. In this case, the intensive blasting induces structural transformations in particles, which favor leaching solution seepage and contribute to nano- and micro-aggregation of dispersed metal fines.

This technology was first time trialed at Talatui deposit within Darasun ore field. Talatui gold–quartz–sulfide–magnetite deposit is composed of veins and interlacing veins and features complex material constitution of ore with dispersed and finely disseminated gold particles.

Based on sampling of muck in blast holes, the open pit mine field is divided into zones: rich zone with gold content more than 2 g/t , low-grade ore zone with gold content of 1–1.9 g/t and off-grade ore zone with gold content of 0.2–0.9 g/t. In 2010–2012, the authors tested the low-grade and off-grade ore from Talatui open pit mine from the viewpoint of applicability of heap leaching of dispersed gold using standard cyanide solutions and photo-electrochemically activated solutions. The tests showed that standard cyanide solutions were inefficient in heap leaching of blasted ore ground down to –100 mm (recovery was under 30%, whereas activated solutions allowed recovery not less than 65% and even more than 80% of gold after fine grinding of blasted ore (–3.5 mm). In the ore block subject to the test blasting, as agreed upon with the mine director (A. Burkser, 2012), the blasting pattern was 1.5×1.5 m instead of 2×2 m (without identification of a transition zone at that stage). After blasting, ore was sent to blending and processing in the circuit of flotation and adsorptive leaching of flotation concentrate. It appeared impossible to separately process ore from the test block. However, at the stage of treatment of blended ore mass, containing an ore portion from the test block, at Darasun processing plant, gold recovery essentially grew (not less than by 15%).

References

[1] Sekisov AG, Zykov NV and Korolev VS 2012 *Dispersed Gold. Geological and Technological Aspects* Mosow: Nedra (in Russian)
[2] Sekosv AG, Zykov N.V., Khakulov VA et al *Preparation of Gold-Bearing Ore to Leaching by Blasting* (in Russian)
[3] Russian Federeation Patent No. 8350896 *Bull. Izobret.* 27.09.2009

Preparation of ore blocks for mine leaching by reagent explosion injection

YuS Shevchenko

Chita Division of the Chinakal Institute of Mining, Siberian Branch, Russian Academy of Sciences, Chita, Russia

E-mail: leshii.ru@list.ru

Abstract. The current drilling-and-blasting operations fail to prepare intact ore body underlying a production horizon for subsequent mining and leaching. It is found that the required preparation quality is possible by means of advanced implementation of ore body discontinuity and filling of the resultant system of joints with active leaching solutions.

Conventional work processes, e.g. drilling and blasting, are oriented in line of advance of the current production face and are therefore inefficient in preparation of underlying intact ore mass for extraction and leaching.

With a view to enhancing mine efficiency by means of reduction in energy input of a technology cycle and owing to an increase in extraction of valuable components, it is proposed to carry out explosion injection as an advanced preparation of an underlying ore block during drilling-and-blasting on a production level [1]. This article presents details of this synergetic approach. After muck has been removed from the face area on the current production level, the underlying block is pre-split by blasting for saturation of the created cracks with leaching solutions. With this end in view, a system of blast holes is made in the underlying block by ring drilling, and a package charge composed of an explosive and capsules with leaching solutions are placed there and initiated, which allows injection of leaching reagents in the pre-cracked zone and makes the period of rock mass relaxation go longer.

Then, drilling-and-blasting is continued at the current production face. Pressure created by the next blasting cycle to advance the production face facilitates deeper penetration of leaching reagents in the preliminary created zone of jointing in the underlying block and prevents cracks and other defects from closure. Such activation of underlying ore mass with leaching reagents weakens bonding between ore and rock particles. As a result, propagation of jointing and penetration of leaching agents proceed both along the current production level advance and towards the underlying ore block (see the figure below).

Saturation of underlying ore by injection during heading operations on the overlying level allows using explosives of lower energy later on. Furthermore, shattered and pre-split ore mass activated with leaching reagents is easier to be processed in the latter cycles of pre-treatment, separation, concentration and leaching.

The application of the proposed technology is described below in terms of Talatui gold-bearing ore deposit within Darasun ore field. The deposit is composed of rebellious high-sulfide ore (pyrite, arsenopyrite, sulfosalt). Gold occurs as visible particles in quartz veins or as dispersed fines in matrix minerals, mostly, quartz and tourmaline. In view of gold dispersion, it is efficient to extract it from mineral matrixes by leaching. It is advisable to prepare rebellious ore for further processing both on the current level and on the lower lying level.

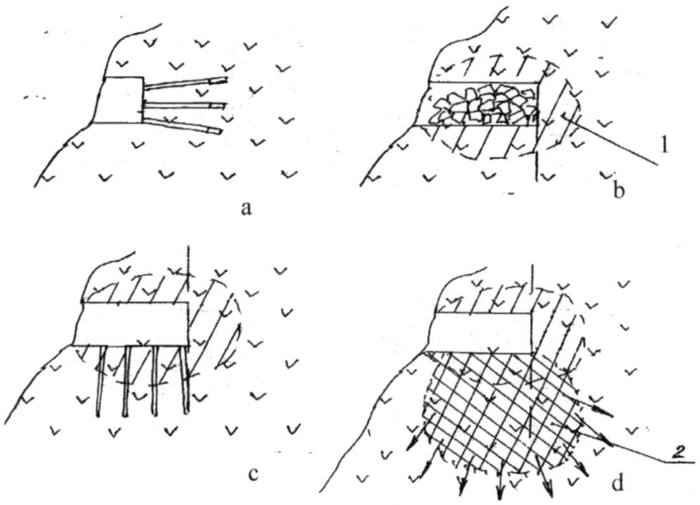

Work cycle: (a) drilling and charging of blast holes at the current production face; (b) face area on horizontal level after drilling-and-blasting (1—pre-splitting zone in the ore block to be activated later on); (c) muck removal and drilling of vertical or inclined blast holes toward the underlying ore block, and installation of charges of explosives and reagent solutions; (d) initiation of the explosive charges and formation of cracks 2. Arrows show direction of permeation of leaching reagents.

When muck is removed from the current production face after pervious cycle blasting, ring and fan drilling is conducted to make blast holes 250 mm in diameter and to charge them with an explosive (ammonite ZhV no. 6, 400 kg per blast hole) added with two capsules. For gold extraction, one capsule contains sodium hypochlorite and the other capsule contains chlorohydric acid. Initiation of the explosive charge breaks the capsules, and the reagents intermingle and exothermally interact (heat emission). Concurrently, blasting disintegrates ore, a damage zone is formed as induced jointing, and the mix of the leaching agents is injected in the joints under action of gases after explosion. Reaction between the reagents generates hypochloric acid that is a key oxidizer for sulfide minerals–matrixes containing gold. In this case, after blasting, leaching process is oriented in the line of the current horizontal production face advance. After blasting, muck is removed from the production face area and is sent for further processing by pre-treatment, radiometric separation, etc. In the clean floor at the production face, vertical holes or, if required depending on the initial jointing of the rock mass, inclined holes with the diameter of 50 mm and 5 m long are drilled with the pattern 1×1 m. In these holes, charges of ammonite ZhV no. 6 (powder factor 1 kg/m^3) are installed with the capsules of sodium hypochlorite and chlorohydric acid, i.e. the same reagents as in the main blasting cycle on the production level. The powder factor of the explosive charge ensures generation of the induced jointing zone with joints oriented vertically or at an angle downward as the blast holes.

In this case, initiation of the explosive charges induces the same events as in the previous cycle on the production level. As a result, concentration zones of induced jointing are formed in the underlying ore block, and leaching reagents penetrate these joints, weaken the ore mass and simultaneously liberate gold from mineral matrixes.

After that, the next drilling and blasting cycle is carried out for further advance of the production face horizontally. This process creates pressure that affects the induced jointing zone in the underlying ore block and contributes to penetration of leaching reagents to deeper levels. As a result, propagation of jointing and penetration of leaching reagents proceed both along the advance of the current production face and downward, toward the underlying ore block meant for mining later on.

Thus, as the production face is advanced, a damage zone of induced jointing is created under it, and free surfaces in these joints are pre-activated with leaching reagents. Owing to this, when mining is transferred to the lower lying level, rock mass there is weakened and less resistant to blasting, which allows using lower energy explosive charges and reduces period of subsequent leaching of valuable components from ore. Also, energy and finance expenditures per mining cycle are cut down. Moreover, it is expected that dynamic hazards (e.g. spalling under high stress) will abate when the lower level face catches up with the upper lying face.

Thus, injection of reagent solutions in ore during blasting allows preventing closure of micro-crack generated by previous blasting, and, consequently, betters further saturation of shattered block with leaching solutions.

References

[1] Shevchenko YuS and Sekisov AG 2015 *Method of Ore Extraction and Preparation for Leaching* Favorable Decision of Application No 2015118529/03(028767)

Relaxation phenomena and jointing in ore bodies under explosive injection treatment

YuS Shevchenko

Chita Division of the Chinakal Institute of Mining, Siberian Branch, Russian Academy of Sciences, Chita, Russia

E-mail: leshii.ru@list.ru

Abstract. Under analysis are relaxation phenomena in ore bodies under leaching. It is shown that liquid and gas–liquid phases make these phenomena last longer, thus it is recommended to apply synergetically interconnected processes of explosive fracture of an ore block and simultaneous injection of active leaching solutions in the generated system of joints.

In a geotechnology cycle, leaching solutions effectively interact with gold micro-aggregates when the latter are separated from the host matrix. To extract gold particles "embedded" in lattices of mineral matrixes, it is required to induce large-scale events of enlargement, diffusion and relocations in crystal lattice, etc. in order to direct migration of ions and atoms of gold [1]. This is feasible with a developed system of defects and concentration of potentials on them. Actually, beneficial effect of jointing system generation and activation of initial defect structure changes into negative relaxation events (R-events) of closure of cracks, relocation of other defects, mineral structure rearrangement, etc. after technological loading is terminated. Considering this, technological stressing should be not a transient narrow-range effect but a volume impact of energy injection to prevent from negative R-events in a system, either solid or dispersed [2].

Some publications, e.g. [3], emphasize the importance of knowing rock mass behavior under external impact and the entire rock mass condition. Rock response to an external impact may be various and show itself as variations of stresses, or induced electrical, thermal and other fields.

Relaxation events are mostly considered from the viewpoint of mechanics of processes running in rock mass [4, 5]. R-events occur on micro- and macro-levels. Inasmuch as residual stresses accumulate when R-events do not occur, it may be stated that macro-scale critical moments are induced in activated systems by small-scale dynamics according to the conclusion drawn based on the R-model in [6].

Closure of old and fresh defects under explosion load is explained in [7] by concentration of opposite charges on surfaces of cracks, while density of the excess surface charge can exceed 10^{10} of electrons per square centimeter (by the data courtesy of Kornfeld [7]). In this case, "dry" cracks are meant. In the presence of liquid and gas–liquid phases, it should be expected that potential R-events will last longer. Such R-events can be estimated based on the data on electromagnetic emission (e.g. [5]) or by variations in electromagnetic induction. For instance, the case study of experimental explosion in gallery no. 13 in Kommunar Mine, Khakassia [7] shows that the record of electromagnetic emission had three maximums: the first was at the time of the explosion, the other two maximums were recorded 0.55–0.6 and 0.83–0.85 min after the explosion and, most probably, conformed with the closure of most energy-saturated newly created cracks. Taking into account phases of scattering of these maximums, it is assumable that the duration of these relaxation events actually was at least a few minutes (the mentioned study commented that delay of post-blasting events was a few minutes after weak explosions (32 kg) and up to half an hour after strong explosions with explosive charges up to a few tons in weight).

In the presence of liquid and gas–liquid phases in a crack, the crack closure is prevented by electro-thermal- and barocapillary events [8] conditioned by deceleration of diffusion relaxation of induced charges. In addition, influence of hydrodynamic factors should be taken into account: liquid phase fills a crack, coats its rough surfaces and viscously resists relaxation closure of the crack [9]. When relaxation processes run in gas–liquid media, in accordance with [10], geometrical similarity of explosion waves, typical of single phase systems (and media), is absent, and it is observed that shock wave parameters strongly depend on completion of relaxation processes.

In a volume activated by blasting, processes associated with the phenomena of heat transmission deceleration take place [11, 12]. Thus, injection of an active reagent in rock mass with a system of growing joints will prolong period of closure of cracks up to tens minutes and even a few hours. This conclusion is very important for application of geotechnologies based on synchronization of external impact on a mineral systems and injection of leaching reagents in cracks in this system.

We assume that defect structure, system of joints, in our case, is primarily in the initial equilibrium. Closure of cracks is the secondary inequilibrium process of relaxation nature. Evidently, closure of one crack takes much less time than closure of zone of jointing spread (in a zone activated by blasting, growing defects emit elastic pulses and initiate a relaxation phenomenon of self-synchronization of defect formation without additional external impact [13]. Naturally, this phenomenon will intensify under injection of gas, liquid or gas–liquid phases in the defect structure).

By analogy with the expression from [14], the approximated time period t for an activated zone to reach equilibrium is given by

$$t \approx (Pz/Pc)^k \tau >> \tau, \qquad (1)$$

where Pz and Pc are dimensions of the activated zone and a crack, respectively, m; k is the characteristic of the medium where relaxation takes place ($k = 1$ for a solid, $k = 2$ for fluid and gas); τ is the time of relaxation (closure) of one crack, s.

The calculations from (1) at Pz = 0.5, 1, 2, 3, 5, 10 m and Pc = 0.05, 0.1, 0.5 m prove that relaxation depends on dimensions of an activated zone as was highlighted in [14]. It is possible to arrive at the same conclusion in a different way. Let in the R-model, the distance between the masses m_A and m_B is l. In this case, the elasticity of the connecting part is $c = E/l$, where E is Young's modulus. Under the impact n, it is possible that

$$\tau = 2\pi/n = 2\pi/(c/m)^{1/2},$$

but $m = pV$, $V = \pi l^3/6$, therefore

$$\tau = 2\pi l^2/(6E/p\pi)^{1/2} = 2\pi l^2(0.52p/E)^{1/2}$$

or, since $p/E = \text{const}$

$$\tau = \mu^* l^2 \qquad | \ \mu^* = 2\pi(0.52p/E)^{1/2}. \qquad (2)$$

Even with the assumed equality of strain wave velocity and P-wave velocity v_P, for most ores where v_P = (0.2–8)×10 m/s, the relation time τ at d = 10^3–10^0 m ($d \sim l^2$) varies from 63.8×10^{-9} to 1.57×10^{-3} s (wave decay is neglected). With a view to defect structure, loading rate and dimension parameter of an activated zone (d), the expression (2) is re-written for real objects in a higher approximation

$$\tau = k\pi d^n t_\varpi / l^* v_P t_{crit},$$

where k is an experimentally obtained coefficient of proportionality.

In the above formula, the n -powder dependence between the relaxation time of stresses (acquired potentials) and the dimension of a system (activated body) means that, as compared with the traditional technologies, activation of a material with regard to R-phenomena is, figuratively speaking, of a submicroscopic nature since the zone of action of the external impact reduces to the dimension of the defects. Finally, we have that velocity of relaxation loss of a potential acquired by a system during activation depends on the number of closing defects (cracks).

These circumstances arising from R-phenomena predetermine the requirement of using synergetically interrelated processes of explosive rupture of an ore block and simultaneous injection of active leaching solution in the blast-generated system of joints.

References

[1] Shevchenko YuS, Sekisov AG and Manzyrev DV 2012 Mechanism of anomalous behavior of gold during blasting and ore pretreatment *Vestn. Zabaikal. Gorn. Koll.*, No. 5
[2] Shevchenko YuS 2014 Relaxation phenomena and blasting injection preparation of ore for underground leaching *FPVGN* No. 1
[3] Egorov PV, Shevelev YuA, Matveev IF, Sklyar N.I and Kvochin VA 1999 *Ground Control in Mines in Gornaya Shoria* Kemerovo: Inst. Uglya SO RAN (in Russian)
[4] Gurevich GI 1974 *Deformability of Media and Seismic Wave Porpagation* Moscow (in Russian)
[5] Bespalko AA Yavorovich LV and Viitman EE 2010 Dynamoelectric energy transfers in a rock mass under explosion load in terms of the Tashtagol Mine *J. Min. Sci.* 46(2) pp. 136–142
[6] Shevchenko YuS 1991 Rock failure model, considering relaxation *Chemistry and Tehcnology of Minerals* Ulan-Ude: BNTs SO AN SSSR (in Russian)
[7] Vorobiev AA, Tonkonogov MI and Veksler JA 1972 *Theoretical Questions of Physics of Rocks* Moscow: Nedra (in Russian)
[8] Balasanyan K 1961 *Dynamic Geoelectrics* Novosibirsk: Nauka (in Russian)
[9] Terzaghi K 1943 *Theoretical Soil Mechanics* John Wiley and Sons
[10] Palamarchuk BI http://www.dissercat.com/content/dinamika-udarno-volnovykh-protsessov-pri-vzryve-v-gazozhidkostnykh-sredakh.html (in Russian)
[11] Novikov IK and Voskresensky KD 1977 *Applied Thermodynamics and Heat Transmission* Moscow: Atomizdat (in Russian)
[12] Kartashov EM 1979 *Analytical Methods in Heat Conduction in Solids* Moscow: Vyssh. Shkola
[13] Manzhikov BTs 2007 Self-synchronization of acoustic emission in rocks under relaxation *Proc. Geodynamics and Stress State of the Earth's Interior* Novosibirsk: IGD SO RAN (in Russian)
[14] Ponomarev VS 1985 Paradoxes of stress relaxation in rocks *Priroda* No. 5

Evaluation and development of integrated technology of rare metal concentrate production in high-level ore processing at Zashikhinsk deposit

MS Khokhulya[1] TN Mukhina[1] VA Ivanova[1] G V Mitrofanova[1] A V Fomin[1] and VD Sokolov[2]

[1]Mining Institute, Kola Science Center, Russian Academy of Sciences, Apatity, Russia
[2]Tekhnoinvest Alliance, Moscow, Russia

E-mail: [1]root@goi.kolasc.net.ru, [2]sokolov@columbite.ru

Abstract. The authors discuss material constitution of columbite ore sample and recommend optimized pretreatment modes to obtain ball milling products at the maximum dissociation of ore minerals in aggregates. A concentration technology is proposed, with division of material into two flows –0.315 mm and –0.2 mm in sizes, generated in the milling and screening cycles and subjected to gravity–magnetic and magnetic–gravity treatment, respectively. It is shown that the technology ensures production of both tantalum–niobium and zircon concentrates. It has become possible to additionally recover rare metal components Nb_2O_5 and ZrO_2 from tailings through flotation.

In the recent decades, an increasing number of international industries use rare metals and rare earths, ore their alloys and compounds. These are, primarily, tantalum, niobium and zirconium, as well as many rare earth elements extracted from mineral concentrates of various tantalum–niobium ore processing. In particular, there are more than 50 known minerals of niobium and minerals, mostly represented by complex oxides [1].

As a rule, such ores contain a number of components concentrations of which justify feasibility and economic efficiency of industrial utilization [2]. Thus, development of a technology for integrated and deep concentration is in this case a critical task from the viewpoints of economic efficiency and environmental safety.

Russia takes one of the world's top leadership positions in terms of proven reserves of most rare metals and rare earths though stands down in terms of quality of these ores characterized by complex texture and structure, low or variable content of useful elements as well as regarding geographical location of such ore deposits.

Russia's reserves of rare earths in terms of oxides make round 28 Mt [3], which ranks Russia second to China in this respect. More than two thirds of Russia's resources of rare earths occur and are under development in the Murmansk Region: unique Khibiny apatite–nepheline field, which holds 42.2% of registered economic resource base of Russia, and the largest Lovozero Ti–TR–Nb–Ta deposit holding 25.4% of reserves [4].

Processing of tantalum–niobium and zirconium ores currently uses conventional pre-treatment followed by sufficiently developed stage-wise flow charts of combinations of gravity, magnetic and electrical separation conditioned by high density and paramagnetic and diamagnetic properties of valuable elements [5, 6]. At the same time, the use of these process results in high metal loss in tailings due to fineness of mineral grains.

Production of niobium and tantalum in Russia is based on loparite ore of Lovozero deposit. The bulk concentrate contains round 95% of loparite including 9% of niobium and tantalum oxides, more than 38% of titanium dioxide and over 32% of rare earths.

As a rare metal supply, Russia plans to use columbite ore of Zashikhinsk deposit to be developed in the East Siberia and source tantalum–niobium and zirconium production. A sample of this ore was taken for technological study to substantiate and develop an integrated processing scheme for this kind of ore. The data on mineral composition and chemistry of columbite ore are given in Table 1.

Table 1. Mineral and chemical composition of columbite ore.

Minerals	Content, %	Elements	Content, %
Columbite	0.30	$\mathbf{Nb_2O_5}$	0.30
Zirconium	0.92	$\mathbf{Ta_2O_5}$	0.026
Xenotime	0.02	$\mathbf{ZrO_2}$	0.59
Thorite	0.01	SiO_2	0.01
Cryolite	0.32	Al_2O_3	0.32
Siderite	0.34	Na_2O	0.34
Amphiboles	0.14	K_2O	0.14
Feldspar	75.52	Fe_2O_3	1.53
Quartz	17.57	CaO	1.16
Mica	4.41	MgO	0.12
Fluorite	0.20	MnO	0.11
Iron oxides	0.18	TiO_2	0.065
Other	0.07	V_2O_5	0.012
Total	100.0	Cr_2O_3	0.0012

The sample contains more than 93% of quartz and feldspar. Mica is represented by muscovite and biotite in total amount of more than 4%. Overall content of rock-forming minerals is above 99%. Fraction of rare metal minerals makes no more that 1.2%. The rare metals are columbite and zircon. TR are represented by xenotime and thorite. Basic components of the ore are silicon oxide and alumina. There overall content exceeds 86%. The ore sample features low content of CaO, MgO, MnO and TiO_2 (0.065%) and insignificant content of Fe_2O_3 (1.53%). Total content of alkalis is 8% with the domination of sodium over potassium. The useful components are Nb_2O_5—0.30%, Ta_2O_5—0.026% and ZrO_2—0.59%.

The optical studies show that columbite occurs in free condition only in fine particles 0.2–0.16 mm in size. This fact is to be taken into account in development of the targeted processing technology for the bulk of tantalum–niobium minerals in larger particles are closely intergrown with the rock-forming minerals. Full dissociation of columbite and aggregates takes place in particles smaller than 0.1 mm.

Extraction of free columbite grains from large size grades of finely milled ore was achieved with the selected optimized mode of ball milling in a closed loop with a screen, which allowed production of screen undersize of -0.315 mm for gravity separation.

The figure below depicts a basic flow chart of gravity-and-magnetic separation of milled ore. Spiral separation of screen undersize with the subsequent table concentration of the rough concentrate produced a concentrate containing 12% of Nb_2O_5, 1.13% of Ta_2O_5 and 23.4% of ZrO_2 to be sent to magnetic separation in a weak field with the magnetic induction not higher than 0.1 T to separate scrap material.

For re-extraction of useful components, spiral separation and table concentration tailings containing aggregates of columbite and rock-forming minerals were mixed and forwarded to ball re-milling down to the size of -0.2 mm.

The magnetic separation of the screen undersize in the low induction field produced scrap material and nonmagnetic fraction composed of minerals with low paramagnetic properties (columbite, tantalite, siderite) and diamagnetic properties (zircon, pyrochlore, quartz, feldspar). It was found that the nonmagnetic fraction subjected to high-intensive magnetic separation with the field induction of 1.5 T was separated into nonmagnetic, intermediate and magnetic fractions.

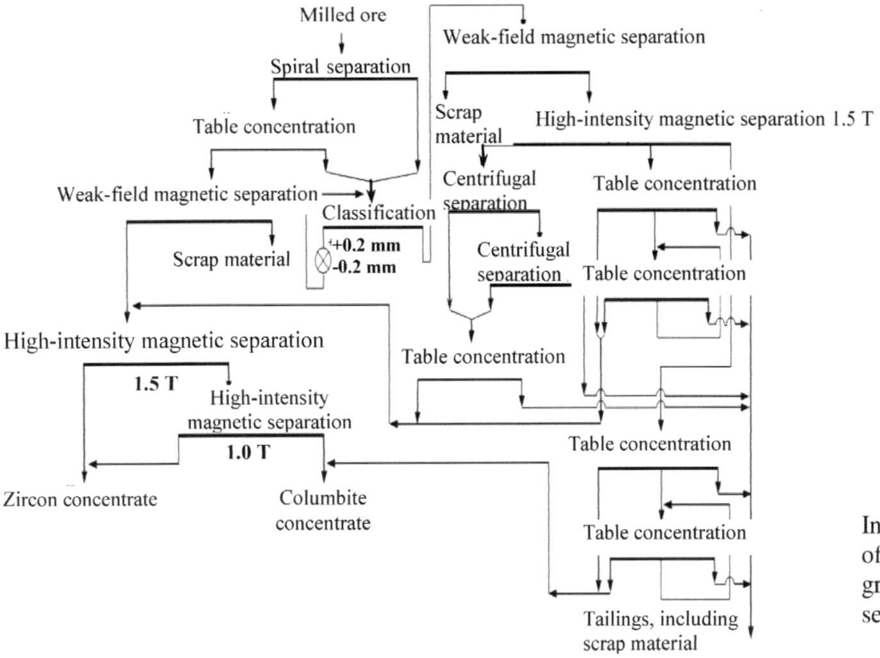

Integrated processing of columbite ore by gravity and magnetic separation.

The subsequent two-cycle centrifugal separation of the nonmagnetic fraction with the rougher and recleaner stages allows a crude product sent to the table concentration. The centrifugal separation and table concentration tailings are discharged (see the figure).

With the aim to improve concentrations of niobium and tantalum oxides and zirconium dioxide, the middlings of the high-intensity magnetic separation are sent to the table concentration to obtain a concentrate containing 12% of Nb_2O_5, 1.13% of Ta_2O_5 and 23.4% of ZrO_2. The middlings are subjected to recleaning and the resultant middlings are mixed with the rougher table concentration middlings. The recleaning concentrate is mixed with the bulk table concentrate to be then subjected to the high-intensity magnetic separation.

The magnetic fraction of the high-intensity magnetic separation goes to table concentration to reach the wanted content of Nb_2O_5 and Ta_2O_5. The table concentration middlings of this stage are fed to recleaning in the next-stage table concentration tailings of which are discharged. The gross concentrate of all table concentrations is a finished product with the overall yield of 0.088%, containing 35.81% of niobium oxide, 2.41% of tantalum oxide and 7.08% of zirconium oxide with the respective recoveries of 10.23%, 9.2% and 1.06%.

The aftertreatment of the combined products represented by rougher table concentrates of nonmagnetic fraction and high-intensity magnetic separation middlings as well as of nonmagnetic fraction of low-intensity magnetic separation after separation of the bulk rougher table concentrate from the spiral separator feed, involves high-intensity magnetic separation under magnetic field induction of 1.5 T. The first stage of this magnetic separation produces nonmagnetic and magnetic fractions. The nonmagnetic fraction is zircon concentrate, and the magnetic fraction is subjected to the second stage separation under the magnetic field induction of 1.0 T, the product of which is columbite concentrate.

In this manner, the developed flow chart of integrated gravity and magnetic separation produces both columbite and zircon concentrates.

Total yield of columbite concentrate made 0.68% with the content of 33.91% of Nb_2O_5, 2.18% of Ta_2O_5 and 10.8% of ZrO_2 at the respective recoveries of 74.83%, 63.56% and 12.48%. The yield of zircon concentrate was 0.835% with the content of 46.41% of zirconium dioxide and admixtures of niobium and tantalum oxides in amount of 2.18% and 0.274%, respectively. The end-to-end recovery of the listed useful components made 65.87%, 5.92% and 9.83%, respectively.

With the aim to additionally extract fine dissociated grains of tantalum–niobium minerals from discharged tailings where loss of Nb_2O_5 and ZrO_2 reached 20% and Ta_2O_5—more than 26%, feasibility of flotation of these elements was studied.

The grain-size analysis of the tailings showed relatively high content of size grade under 0.071 mm (40%), with the concentrations of Nb_2O_5 and ZrO_2 round 67% and more than 60%, respectively. Niobium oxide content of the tailings was 0.065% and zirconium dioxide—0.098%.

The mineralogical analysis showed the need for additional milling which increased the content of the size grade -0.071 mm up to 64.3%, with the higher content of free grains with renewed and activated surfaces.

Prior to flotation, the tailings were deslimed to remove size grade 20 μm in an open cycle with fresh water; then subsequent cycle contained stages of the rougher flotation (RF) and recleaner flotation (ReF) and 2–3 stages of scavenging of froth product of RF (S-1, S-2 and S-3).

The tests used the known reagent regimes in alkaline medium (pH approx. 10) [7–9] with the collecting agent represented by hydroxamic acid (HA) in combination with fatty acids and mines, or xanthate with the frother T-80 and liquid glass as a depressor. Another reagent regime tested was a collecting agent from the class of esters of phosphorus-berating compound in combination with sulphoxyl reagent and finishing of RF froth product in strong acid medium (pH = 2) [10]. However, with the use of these reagent modes, the concentration of useful components in RF froth product was low—6–14 at the recovery no more than 60%. Scavenging stages, either without reagents or with ethane diacid, were inefficient. The bulk concentrates contained 2.71% of ZrO_2 and 1.65% of Nb_2O_5 at the recovery of 52.8% and 48.2%, respectively, at the rate of concentration not higher than 27 and 25.

For the metals of Zr, Nb and Ta are highly prone to complexing, the selected collecting agent was alkyd hydroxamic acids with a high constant of complexing, including zirconium ions [11]. The selected agent HA was analogous to the reagent IM-50—a mix of 75% of alkyl (C_7–C_8) hydroxamic and 25% of carboxylic acids at a ratio of 3:1.

The use of HA in combination with different depressing agents for rock-forming minerals, such as sodium hexametaphosphate (SXP), liquid glass and sodium lignosulfonate, which were at the same time dispensers for slime particles, exhibited the highest efficiency of SGP.

Table 2 reports the results of flotation with HA (pH = 9.7) after pulp slurry treatment with SGP.

The froth product contains 65.7% of Nb_2O_5 and 73.5% of ZrO_2 with the sufficiently lean tailings (0.019% of Nb_2O_5 and 0.016% of ZrO_2) with weak concentration rate (4); the effect of selective separation shows in the scavenging operations that yield concentrates S-1 and S-3 with the contents of 4.67 and 7.42% of Nb_2O_5 and 7.98 and 13.02% of ZrO_2 at the recovery of 61.8 and 50.2 and 70.0 and 548.5% and the rates of concentration of 130 and 115, respectively.

The reagent regime with SXP and HA agents was tested in flotation of milled tailings without desliming. The results (Table 2) show that at the optimal consumption of the reagents, it is possible to obtain the bulk rare metal concentrate with the content of 6.78 % of Nb_2O_5 at the recovery of 58.4% and 12.28% of ZrO_2 with the recovery of 70.2%. The optimal pH ranges as 9.7–10.0 with SPX and HA. When pH is under 9.7, floatability of the collecting agent drops; when pH exceeds 10 floatability grows, which leads to the need for additional scavenging stages for RF froth product.

Conclusion

The analyses of the material constitution of a columbite ore sample containing 0.3% Nb_2O_5, 0.027% Ta_2O_5 and 0.526% ZrO_2 have shown that the basic mineral of the ore are columbite, zircon, quartz and feldspar.

Table 2. Flotation of tailings of gravity and magnetic concentration of columbite ore.

Dressing product	Results, %					Test conditions and reagent consumption, g/t
	γ	Nb_2O_5		ZrO_2		
		β	ε	β	ε	
Pulp slurry	11.1	0.066	11.3	0.092	10.4	With desliming;
RF froth product	18.4	0.232	65.7	0.392	73.5	**RF:** SXP—900
ReF froth product	7.9	0.042	5.1	0.075	6.0	HA—1350
Tailings	61.6	0.019	17.9	0.016	10.1	ReF: HA—225
Initial feed	100.0	0.065	100.0	0.098	100.0	
S-2 concentrate	0.86	4.67	61.8	7.98	70.0	
S-3 concentrate	0.44	7.42	50.2	13.02	58.5	
RF froth product	29.5	0.172	78.0	0.288	86.4	Without desliming
ReF froth product	15.4	0.032	7.6	0.036	5.7	**RF**: SXP—1000
Tailings	55.1	0.017	14.4	0.014	7.9	HA—1670
Initial feed	100.0	0.065	100.0	0.098	100	ReF: HA—300
S-2 concentrate	1.10	4.14	70.1	6.88	77.3	
S-3 concentrate	0.59	6.58	59.8	11.48	69.1	
RF froth product	5.6	0.889	76.0	1.559	88.4	Without desliming
Tailings	94.4	0.017	24.0	0.012	11.6	**RF**: SXP—2000
Initial feed	100.0	0.065	100.0	0.098	100.0	HA—2670
S-1 concentrate	1.0	4.43	67.6	7.939	81.0	
S-d concentrate	0.56	6.78	58.4	12.28	70.2	

The authors have selected optimal modes for ball milling of ore and dressing products to ensure high dissociation of columbite grains, which enables preparing screen undersize for subsequent processing cycles.

The developed technology of combination gravity–magnetic concentration of columbite ore ensures production of tantalum–niobium and zircon concentrates:

—Columbite concentrate: yield: 0.68%;
content: 33.91% Nb_2O_5, 2.18% Ta_2O_5 and 10.8% ZrO_2;
recovery: 74.8%, 63.6% and 12.5%, respectively.

—Zircon concentrate: yield: 0.835%;
content: 46.41 % ZrO_2; 21.8% Nb_2O_5 and 0.274% Ta_2O_5;
recovery: 65.9%, 5.9% and 9.8%, respectively.

The studies have demonstrated feasibility of improving completeness of extraction of valuable components by means of concentration of tailings based on flotation and an optimal reagent regime in an alkaline medium using a complexing collecting agent represented by hydroxamic acids and sodium hexametaphosphate as dispergator of slime particles and depressor of barren rocks. The resultant open-cycle bulk rare metal concentrates contain 6.78% of Nb_2O_5 and 12.28 % of ZrO_2 at the recovery of 58.4% and 70.2%, respectively.

References

[1] Korovin SS, Drobot DV abd Fedorov PI 1999 *Rare and Dispersed Elements. Chemistry and Technology: University Textbook* Moscow: MISIS (in Russian)
[2] Kurkov AV and Anufrieva SI 2014 Technological issues of intergated processing of rare

metak–rare earth oores and their solutions *Plaksin's Lectures 2014: Int. Conf. Proc.* Almaty: ARKO (in Russian)
[3] Rubanov IA 201 Basic elements *Ekspert* No 44
[4] Mikhailov BK, Kiperman YuA, Komarov MA et al 2012 *Minera Resources of Industrial Waste* Moscow: Nauch. Mir (in Russian)
[5] Polkin SI 1987 *Concentrateion of Ores and Rare and Precsious Metals* Moscow: Nedra (in Russian)
[6] Jordens A, Sheridan RS, Rowson NA, and Waters KE 2014 Processing a rare earth mineral deposit using gravity and magnetic separation *Minerals Engineering* Vol 62
[7] Shubov LYa, Ivankov SI, Shchegkova NK 1990 *Flotation Reagents in Mineral Dressing. Book of References* Moscow: Nedra (in Russian)
[8] Wang C-H, Qiu X-Y, Hu Z, Li H-W, Wang T, and Zou J-J 2012 Study on improved benefication technology for rare earth ores *International Mineral Processing Congress* New Delhi, India
[9] Ivankov SI, Lyubimova EI, Lifirenko VE, abd Chuvaldzhyan AG 2003 RF Patent No. 2220006 Method of finely dispersed niobium ore flotation *Byull. Izobret.* No. 36
[10] Kurkov AV and Pastukhova IV 2010 RF Patent No. 2381073 Method of flotation of rare metals and tin *Byull. Izobret.* No. 4
[11] Ivanova V and Mitrofanova G 2015 Flotation of eudialyte: correlation of experimental data with the results of quantum-chemical calculations *Proc. XVI Balkan Mineral Processing Congress* Belgrade

Environmental pressure reduction with a new method of noble metal recovery

EV Filippova

Transbaikal State University, Chita, Russia

E-mail: filena 78@mail.ru

Abstract. Discoveries in the area of hydrometallurgy of noble metals can be of use in metal recovery from low-grade solutions and slurries, including liquid tailings. Efficiency of noble metal recovery and reduction in mining waste is gained owing to utilization of two forms of ion-exchange sorbent, including OH^- for recovery of cyanic compounds of gold and cyanides, which allows abating burden on natural systems.

Easy gold reserves have almost been depleted so far. Major gold reserves are now composed of very fine particles. Moreover, concentration of gold in ore is low. None of the known beneficiation circuits provide complete recovery of such gold, and some gold is lost in tailings in the form of overground aggregates with gangue. The average gold content of ore is 0.001% (10 g/t), and tailings contain 99.9% of extracted rocks. By 2015 Transbaikalia has accumulated 2.9 Bt of waste in dumps and tailings of flotation and gravity concentration [1]. Industrial waste dramatically impairs the environment, including hydrosphere, pedosphere, flora and fauna. Waste substances, including toxic, spread from tailing ponds with water or when dams are destroyed. In drained areas or from concentrate storage yards, toxic matters go with wind and also invite danger.

For the ecological relaxation and in order to enhance recovery of gold and associate chemical elements that are most often pollutants, new approaches and technologies should be developed considering the environmental safety. In this respect, the research has been undertaken to analyze feasibility of development of an efficient and integrated technology for minimization of mining and processing waste and to regenerate eco-systems in the areas of operating mines and processing plants with regard to their specificity.

The scope of the research embraced theories and generalizations, multi-factor experimentation planning, mathematical processing of the experimental results, granulometric, mineralogical, spectral, chemical, X-ray phase, optical, electron microscopic, microscopic, bacterioscopic, atomic absorption, assaying, X-ray structural and another analyses, process testing, desk studies as well as large-scale laboratory and semi-industrial trial of additional leaching of gold with preparation of feedstock in solutions and pulp slurry.

In the course of the research implementation, the process of noble metal recovery has been improved [2] under the guidance of Dr Eng Sekisov, which enhanced efficiency of recovery of gold, chemical combined elements and activated leaching solutions due to the re-designed electro-absorption plant, specific flow of pulp slurry, two new species of an ion-exchange sorbent and increased activity of adsorption and deposition of noble metals on cathodes.

A distinctive feature of the new method to extract noble metals from pulp slurry is additional milling of mineral particles of pulp slurry to unlock dispersed gold particles before feeding the pulp slurry in the electro-absorption plant. The ion-exchange sorbent is fed in the plant first as CN^-, for additional leaching of dispersed gold at early stages of recovery, and then as OH^-, to extract cyanides and cyanide–gold complexes at later stages of the circuit. This allows mitigation of anthropogenic load on the nature and produces the beneficial ecological effect.

Figure 1 shows the layout (vertical) and A-A cross section of the electro-absorber designed for small-scale industry.

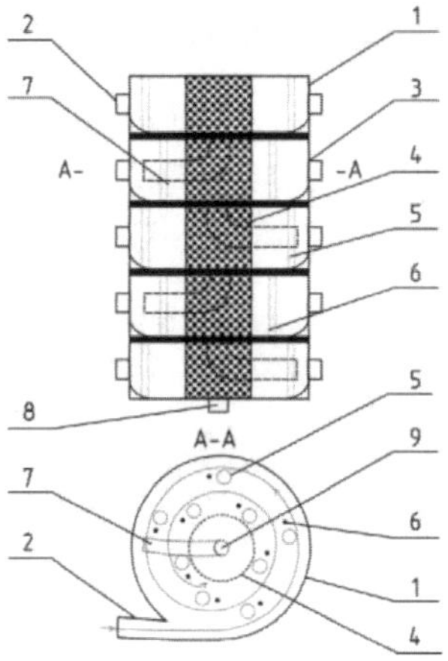

Figure 1. Electro-absorber for the small-scale industry: 1—body composed of five units (external cylinder); 2—inlet pipe; 3—outlet pipe; 4—internal perforated cylinder; 5—cathode; 6–anode; 7—bypass pipe; 8—discharge pipe (to tailings pond); 9—bypass hole.

The body 1 of the electro-absorber contains a number of units with inlet pipes 2 to let in pulp slurry from the pulp slurry pipeline and outlet pipe 3 to let out pregnant ion-exchange sorbent. Inlet pipe 2 is arranged at an angle relative to the inner surface of the external cylinder of the body. The body 1 of each section is composed of two cylinders (external and internal) that are rigidly coupled; the radius of the internal cylinder makes 0.23–0.25 of the radius of the external cylinder. The internal cylinder 4 has perforation. The size of the perforation holes is chosen based on the condition of separation of the ion-exchange sorbent with gold anions and the pulp slurry fluid and its free outflow.

In the external cylinder 1, electrodes 5 and 6 are installed one behind the other along the pulp slurry flow, which reduces abrasive effect on the protective corrosion-resistant surface of the anodes. The difference of the potentials between the anodes and cathodes is conditioned by the physicochemical parameters of pulp slurry and by the content of ion-exchanger in it, and should be sufficient to ensure high-rate electrochemical processes in the electro-absorber.

From hole 9 in the bottom of each unit of the internal perforated cylinder, the bypass pipe 7 goes to the lower lying unit of the external cylinder of the electro-absorber. The bypass pipe is curved to amplify centrifugal force and to direct the pulp slurry flow to the next unit of the plant.

Separation of the pregnant ion-exchange sorbent and discharge of the pulp slurry from a unit takes place in the internal perforated cylinder 4. The pulp slurry to be discharged to a tailings pond is let out via the discharge pipe 8.

Granules of the ion-exchange sorbent are placed in the electro-absorber before the pulp slurry is fed there. In our case, the gold-selective sorbent of grade A-100 was used in amount of 10% of the volume.

Pulp slurry flows comes from the pulp slurry pipelines, through vortex generators installed in the expanded part of the pipeline in front of the inlet pipe for additional grinding of mineral particles containing fine gold.

Then, the flow goes in the external cylinder 1 of the electro-absorber though the inlet pipe 2 where mineral particles collide and bump against the cathodes. The prepared ion-exchange sorbent in the

IOP Conf. Series: Earth and Environmental Science **53** (2017) 012040 doi:10.1088/1755-1315/53/1/012040

form of CN^- mixes with the pulp slurry, interacts with mineral particles, leaches unlocked gold fines and is saturated with them. At the later stage of this circuit, the pulp slurry interacts with the ion-exchange sorbent in the form of OH^-, and additional recovery of cyanides and gold-and-cyanide complexes takes place. This reduces the load on the nature as migration and hypergenesis of cyanides in tailings ponds is eliminated.

The ion-exchange sorbent granules and the pulp slurry are separated when the pulp slurry flows through the perforated holes of the internal cylinder 4. The pulp slurry, after electroactivation and electro-deposition in the first unit, freely flows through the perforation holes of the internal cylinder 4 from the first unit to the second unit, then via the bypass pipe 7—to the external cylinder of each lower-lying unit and, finally, is let out from the electro-absorber by the discharge pipe 8. The ion-exchange sorbent granules, owing to the centrifugal force, circulate in the electro-absorber space between the internal and external cylinders and remain there until complete saturation. The pulp flow rate in a unit ensures the pre-set recovery efficiency.

In the zone between the cathodes and anodes, partial electrolysis takes place with liberation of gas bubbles that favor self-clearing of the perforated surface 4 and active circulation of the pulp slurry and ion-exchange sorbent in the electro-absorber. When saturated, the ion-exchange sorbent is unloaded from the plant and is sent to desorption using some known method.

Figure 2 shows the layout of the electro-absorber intended for application in the large-scale industrial plants.

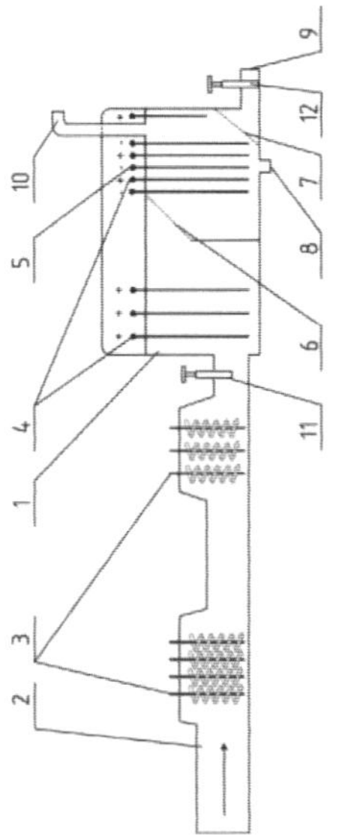

Figure 2. Layout of the large-scale industry electro-absorber: 1—body of the electro-absorber; 2—pulp slurry pipeline; 3—vortex generators; 4—anode; 5—cathode; 6—mesh; 7—mesh; 8—sorbent discharge pipe; 9—discharge pipe; 10—gas outlet pipe; 11 and 12—stop valves.

The second-version electro-absorber is designed for noble metal recovery at large-scale plants. In this case, the ion-exchange sorbent granules are placed in the body 1 before the feed of the pulp slurry. When the pulp slurry flows through the vortex generators 3 installed in the pipeline 2, additional grinding of mineral particles unlocks dispersed gold. Then, in the electro-absorber, mineral particles collide and bump against the cathodes 4, and additional leaching of unlocked gold takes pace. The ion-exchange sorbent in the form of CN^- interacts with the mineral particles, dissolves unlocked gold particles and gets filled with them.

Then, the flow enters the second unit of the electro-absorber through the mesh 6. The mesh catches CN^- sorbent whereas the pulp slurry flows into the second unit with the ion exchange sorbent in the form of OH^- for extra leaching of gold-and-cyanide complexes and toxic cyanides. When passing the anode–cathode system with the constant voltage maintained, destabilized ions go to the ion-exchange sorbent owing to electrodialysis. Oxygen is consumed in the boundary solution layer immediately adjoining gold ions. Further fruitful dissolving of gold is possible if the boundary layer is filled with the ion-exchange sorbent and oxygen from the entire volume of the solution owing to diffusion; the gold dissolving rate will be higher with the higher diffusion rate. Additional oxygen appears on the anode, which favors active leaching and absorption of gold. Gases are discharged via the pipe 10.

As a result, cations go to the cathode, interact with the new particles of the ion-exchange sorbent, generate new polyatomic gold anions and are caught by the sorbent. Or, vice versa, anions go to the anode where additional leaching of gold takes place. Under the combination effect of electro-osmosis and electroactivation of ion exchange, metal ions pass to gel phase of the ion-exchange sorbent. Also, the local ion flows and the associated magnetic microfields additionally intensify the ion-exchange absorption process. Owing to an increase in the local concentrations of ions, some gold ions deposit on the cathode 5 under effect of polarization in the area of the electric layer.

The ion-exchange sorbent OH^- granules are separated from the pulp slurry by the mesh 7. The pulp slurry, after electroactivation and electro-deposition in the second unit, freely flows through the mesh and is let out via the discharge pipe 9 at the flow rate ensuring the pres-set recovery efficiency. As the ion exchange sorbent is saturated, it is unloaded from the electro-absorber and fed to desorption.

The presented technology has been trialed on tailings of rebellious ore concentration at one of the mineral deposits in Transbaikalia. The tailings contain mainly dispersed gold. These gold fines are unrecoverable by leaching using a simple or oxygen-filled cyanide solution or by a regular ion-exchange sorbent. Low gold content of 1.1–1.3 g/t of these tailings prevents from application of any know metallurgical processing techniques.

Multiple trials have proved the efficiency of the proposed technology. Gold recovery has made 57.5%. Reduction in the environmental impact is achieved by application of the ion-exchange sorbent in the form of OH^- to recover toxic gold-and-cyanide complexes and cyanides at the late state of absorption. Furthermore, excess cyanides are oxidized on the anode down to cyanates that are less toxic for the natural environment.

The proposed technology implemented using the designed electro-absorbers favors minimization of waste and mitigation of the environmental impact.

References

[1] *Aggravating Problem Connected with the Local Mining and Processing Industry Waste* Official site of the Public Prosecutions Department of Transbaikalia, Russian Federation. Last visited: 24 November 2015

[2] Sekiosv AG and Mazurkevich SA 2005 RF Patent No 2251582 Noble metal recovery from solutions and pulp slurries and the reaction vessel *Buill. Izobret.* No 13